Kobenan Etienne BINI
Augustin Tiyégbo TOURE
Kouadio Eugène KONAN

Geomorphological problems in non-dam lowlands

Kobenan Etienne BINI
Augustin Tiyégbo TOURE
Kouadio Eugène KONAN

Geomorphological problems in non-dam lowlands

Farmers' adaptation strategies in Katiola, Côte d'Ivoire

ScienciaScripts

Cover image: www.ingimage.com

This book is a translation from the original published under ISBN 978-620-6-70659-5.

Publisher:
Sciencia Scripts
is a trademark of
Dodo Books Indian Ocean Ltd. and OmniScriptum S.R.L publishing group

120 High Road, East Finchley, London, N2 9ED, United Kingdom
Str. Armeneasca 28/1, office 1, Chisinau MD-2012, Republic of Moldova, Europe
Printed at: see last page
ISBN: 978-620-8-15907-8

Contents

DEDICATION

This work is dedicated to my family

FOREWORD

Presented as a complex science, geography is concerned with the study of physical, biological and human phenomena located on the earth's surface. It is therefore interested in their distribution, the factors that organise them and the forces that govern them, as well as their reciprocal relationships. Physical geography, a branch of geographic research, studies the physical foundations of the environment. In its field of critical reflection, it serves to better understand the structure and functioning of natural phenomena and the processes that shape and impact the landscape. On the basis of the explanations it provides, it is possible to put forward practical proposals for improving the living and working environment of society, particularly at the level of a local rural community.

In general, rural communities in the tropics are characterised by various forms of vulnerability to natural hazards. These hazards take the form of permanent or recurrent constraints that cannot be overcome by the population's low level of technology.

It is in relation to this theoretical context of nature-society interaction that we are undertaking a reflection on '***the geomorphological problems of the lowlands managed outside dams and farmers' adaptation strategies: the case of Katiola***'. This is a geographer-physicist's view of the question of societal adaptations to the nature and hydrodynamic functioning of a lowland in north-central Côte d'Ivoire, specifically at Katiola. The socio-spatial context is one of geomorphological interactions that complicate agricultural activities, specifically traditional lowland rice-growing that has not yet benefited from hydro-agricultural dam-type infrastructure.

I PART

GENERAL INTRODUCTION

First part:

GENERAL INTRODUCTION

The aim of this section is to introduce the subject. It situates it in the literary and scientific circles from which the issues raised emerge.

1- INTRODUCTION

1-1- INTEREST OF THE SUBJECT

The economy of most developing African countries is predominantly agricultural, or even agro-pastoral. In the macroeconomic context, only a sound government policy on the management and planning of projects in the essentially agricultural sector, and a high level of public involvement, can boost the development of rural countries.

Although it aspires to become an emerging country, Cote d'Ivoire's economy is undeniably based on agricultural foundations. This justifies the agro-industrial option as a development strategy, with the main structural assets being a young population, a dynamic and cosmopolitan demography, and major biophysical advantages under a climate that is largely favourable to agricultural speculation of all kinds. All of the country's ecological zones would be potential host areas. So it's hardly surprising that Cote d'Ivoire is regarded as one of Africa's agricultural powerhouses.

In terms of international trade and economic globalisation, coffee and cocoa, which have always been listed on the stock exchange, remain one of the country's flagship products. With other agricultural products such as palm oil, hevea, cashew nuts, sweet bananas, pineapples, etc., mining resources such as oil and natural gas, gold and bauxite, and many other minerals, electricity which is produced in abundance - since the country exports it to neighbouring countries - this small country (322,462 km^2) has enjoyed remarkable economic success since 1960. Cote d'Ivoire has consolidated its agricultural leadership through the diversification of agricultural activities, the development of which is based on high-performance agronomic research.

With this in mind, rice is one of the crops promoted by SODERIZ, the state company responsible for developing rice growing, in the 1970s. The challenges remain considerable: rapid urbanisation of the country and strong demographic growth, standardisation of eating habits, since rice has become the main foodstuff for almost all the inhabitants of Côte d'Ivoire. Yet local production still falls short of expectations. This has led to the importation of rice from Asia. Incidentally, the government's current aim is to make Côte d'Ivoire a self-sufficient rice-exporting country. To meet these challenges, we need to combine scientific knowledge, engineering know-how and the voluntary adaptation mechanisms of rice growers, so that the food independence project has an overall impact on the country.

1-2- JUSTIFICATION OF THE SUBJECT

Meeting food requirements is one of man's basic needs. In a national context of rapid population growth, it is important that production approaches and exceeds food requirements at local, regional and national levels. It is therefore necessary to remove all the social and physical obstacles likely to pose a threat to achieving the objectives of increasing agricultural productivity. Rice production still falls far short of expectations, despite the fact that the natural environment is conducive to large-scale promotion of this crop, particularly in central-northern Côte d'Ivoire, a land of plateaux and lowlands with favourable soil conditions.

The specificity of this work, over and above the socio-economic aspect, is to highlight how geomorphological problems can act as a brake on rice production if they are not properly managed by the farmers.

1-3- PRESENTATION OF THE STUDY AREA

The town of Katiola is located in the centre-north of Côte d'Ivoire at 8° 30' North and 5° 10' West at an average altitude of 400 m. It is the capital of the Hambol region. It is surrounded by the towns of Dabakala, 90 km to the east, Mankono to the west, some 166 km away, Niakaramadougou to the north, 63 km away, and Bouake, Botro and Beoumi to the south, respectively 54 km, 41 km and 81 km away. Katiola is also 450 km from Abidjan, the economic capital of Côte d'Ivoire, and 140 km from Yamoussoukro, its political capital (Figure 1).

1-3-1- The physical environment

- The climate

The general climatic context is that of a tropical region characterised by heat and humidity. From a regional point of view, the climate is a tropical transition between that of the semi-arid sub-Saharan zones and that of the humid tropical zones of the Gulf of Guinea. Locally, Katiola is subject to a Sudano-Guinean climate, more humid than that of the far north, with two rainy seasons running from March to June and September to October. These are interspersed with two dry seasons, from November to February and July to August. This climate is controlled by the movement of the Intertropical Front (ITF). Annual rainfall generally varies between 1,100 mm and 1,200 mm, while the average temperature is around 27°C.

- Hydrology

The department of Katiola is crossed by the tributaries of two rivers which form the natural boundaries of the department. These are the N'Zi and the Bandama, to the east and west respectively. It is the latter and its tributaries (Bou, Nabion) that irrigate most of the department. The hydrological regime is of the transitional equatorial type. As for the rivers, their regime is modelled on the tropical rainfall regime.

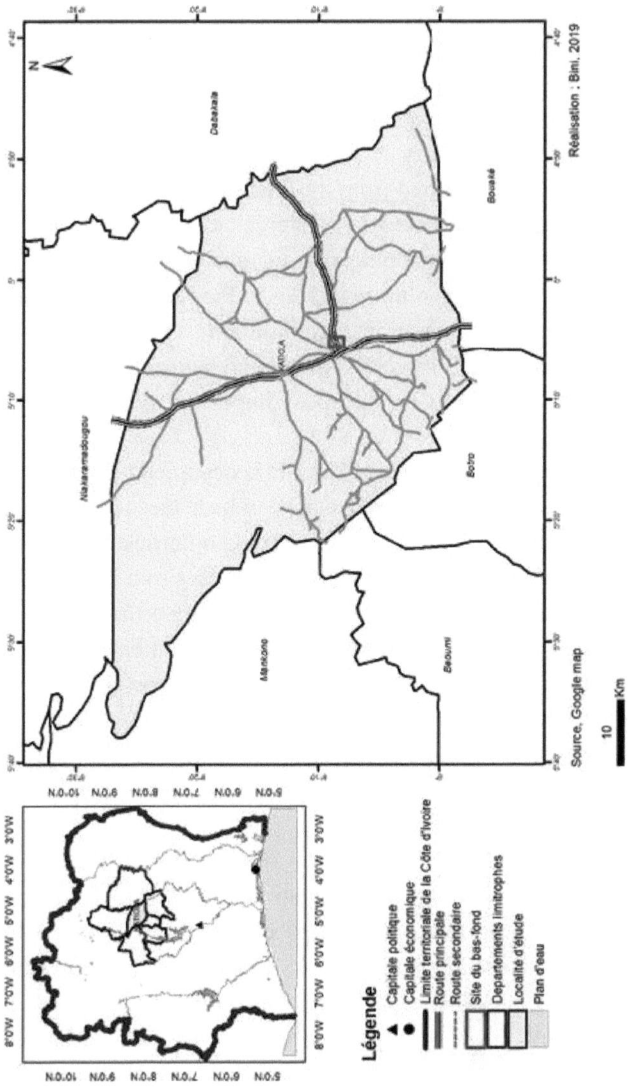

Figure 1: Location of the study site

- Vegetation

The vegetation of the Katiola region belongs to the southern part of the Sudanese domain (Coulibaly, 2009, p.29). More specifically, this is the sub-Sudanese domain, characterised by open forest. The dominant vegetation ranges from shrubby savannah to grassy savannah, with the presence of forest galleries along the watercourses.

Grassy savannahs are most often found on cuirass outcrops at the top of debris plateaus, at the edge of forest galleries or in marecageous lowlands where they constitute hydromorphic savannahs. Hots of lesser importance are found at the edge of paths or roads (Adjanohoun, 1964, p.117).

- Geomorpho-pedology

Soils in Côte d'Ivoire are formed from the alteration of metamorphic bedrock. The soils have not been rejuvenated by tectonic activity; they are old and often strongly altered. Regional mapping carried out using the morpho-pedological approach revealed that the evolution of soils in the Katiola district depends on the following pedogenetic processes: induration or armouring, reworking, pedoplasmation, alteration, impoverishment and hydromorphy (ORSTOM, 1987, p.81). Induration appears to be the most important process in northern Ivory Coast.

The central part of the department of Katiola is occupied by a band of granite with highly heterogeneous and fractured edges, which has led to the formation of numerous inselbergs and areas of whaleback outcrops. To the north-east of Niakaramandougou, the Niangbo granite dome rises to a height of 694m. Like a huge glacis all around, all the other peripheral compartments are much quieter, taking the form of very monotonous plateaux (Coulibaly, 2009, p.31) whose average altitude is slightly above 400 metres, with a drop towards the south-east in the Katiola sector, where the altitude drops to 250 metres.

Because of its savannah cover, this region is highly susceptible to run-off and erosion.

1-3-2- The human environment

- Social framework

The commune of Katiola has a population of 56,681 (RGPH, 2014, p.21). Originally, Katiola was inhabited by two ethnic groups: the Tagbanas and the Mangoros. Immigration has added other Ivorian and foreign communities, in particular West Africans (Malians, Burkinabe, Guineans, Senegalese, etc.) living in perfect harmony. It's a beautiful human mosaic, but with a high sex ratio (107.0%) according to the 2014 census.

Katiola presents a relatively satisfactory picture in terms of basic infrastructure: electrification, water supply, schools, health centres, banks, etc.

In short, Katiola is a modern town. The villages are also undergoing a process of modernisation, with a general trend towards solid construction as part of a similar housing development.

The urban population of Katiola is much higher than the rural population. Clearly, the disproportion between the two populations can be explained by the fact that the departmental capital is home to far more public infrastructure: administration, secondary schools, health facilities, etc.

- Economic life

Roads are a major vector for economic development. The department is crossed by the A2 trunk road, the main south-north route leading to Burkina Faso and Mali. This road also links Katiola to Fronan, Niakaramandougou and Tafire.

In terms of economic activities, the Tagbana rural world is dominated by agriculture. Traditionally, the Tagbana are farmers. Food crops include rice, groundnuts, mats, yams, cassava, millet and sorghum. The main cash crop is cotton, which is now being replaced by cashew nuts. There are also citrus and mango orchards. In terms of handicrafts, the Mangoros potters are widely renowned. Socio-economic activity in the town is driven by civil servants and shopkeepers.

The department of Katiola is characterised by a large-scale migration of young people to other sub-prefecture capitals on the outskirts and to the southern regions of Côte d'Ivoire. Since 2000, the discovery of gold panning sites has concentrated the migratory flow towards them. The environmental impact of this new activity is destructive.

1-4- JUSTIFICATION FOR THE STUDY AREA

Katiola, a town in north-central Côte d'Ivoire, is rich in lowlands. However, previous studies have revealed the low productivity of these lowlands, generally due to socio-cultural reasons. Analyses have concluded that the contribution of these lowland areas to rural development is very mixed (Toure, 1999). But beyond the socio-cultural aspect, other scientific explanations can be put forward.

Among the possible factors, the structural characteristics of the lowlands, including the quality of the micro-models, could have a definite indirect impact on the productivity of these sites. They may therefore constitute major constraints, which, in addition to economic and technical constraints, limit the contribution of inland valleys to the local economy. Other factors include the resilience of the local farming population through greater participation in the design and implementation of cropping methods to counterbalance socio-cultural constraints.

2- ISSUES

2-1- UNDERSTANDING THE SUBJECT

Like any country, Cote d'Ivoire makes the most of the many natural resources at its disposal, given its geographical location in humid West Africa, by exploiting its forest area. Agriculture is one of the pillars of its economy. For a long time now, the agricultural sector alone has accounted for around 40% of gross domestic product between 1990 and 1997, 70% of export earnings and employs 2/3 of the working population (Ministry of Agriculture and Animal Resources, 1999). Although agriculture was originally mainly centred on the coffee-cocoa binomial, more and more room is being made for crops such as rice, a commodity that is widely consumed in the country and which is grown in the lowlands. This speculation is supported at national level by development agencies such as ANADER and CNRA, and by several international partners including the FAO. This pooling of research efforts is boosting yields, enabling Côte d'Ivoire to achieve its objectives, including its goal of self-sufficiency and even food security.

For good reason, it has to be said that the lowlands are under-exploited. This is because most of them are still farmed on a small-scale basis. Not to mention the fact that these environments are particularly difficult to exploit, unlike the interfluves. From a functional point of view, because of their hydrological and sedimentary properties, between which a balance must be found, the lowlands are home to fertile land, but one that is difficult to farm, given the technical context of subsistence agriculture in Côte d'Ivoire.

If we add to this the sad reality of lowland areas that have been developed outside dams, we can see that the lack of control over the technical conditions for irrigated rice growing is a weakness in lowland farming in the country as a whole, which is both cyclical and structural. In particular, the lack of control over water systems is one of the major weaknesses of lowland rice growing, illustrated either by major water surpluses or deficits on farms. This problem is linked to the management of the micro-relief of the plots, which generally suffer from poor levelling. A thorough understanding of the physical and territorial conditions of farming activities is a scientific necessity. It is even a prerequisite for boosting the productivity of the land. This applies to inland valleys, which are presumed to support the agriculture of the future, in the context of wetland management, in the light of the concerns raised by current climate change.

2-2- STATE OF RESEARCH

2-2-1- Lowlands as landscape features

In tropical zones, lowlands make a significant contribution to improving food security. These natural areas, which are analysed in terms of their relationship with society, constitute systems that are an integral part of the landscape, which Bertrand (1978) defines in the following terms :

"Landscape is not the simple addition of disparate geographical elements. It is a certain portion of space, the result of the dynamic, and therefore unstable, combination of physical, biological and human elements which, by interacting dialectically, make the landscape a unique and indissociable whole.

As systems integrating ecological and human dimensions, lowlands reflect the adaptation of a raw natural environment to the needs of society. This is why they are considered to be a product of society.

There is a growing interest in lowland farming systems. There are reasons for this trend. According to Kafilatou (2017, p.1601), if the use of lowlands has increased considerably in number and area, it is because of the fertility of their soils and their hydromorphic character. However, at the soil level, the superficial pedological organisations undergo transformations under the effect of meteorological, faunal and anthropic factors. There are also wind transfers at the surface and water transfers at depth.

The usual typologies of lowlands are drawn up according to a combination of parameters: Albergel et al (1993, pp. 233-235), according to geological and climatic characteristics; Albergel and Claude (1988), according to climatic and geomorphological characteristics, as well as Raunet (1982) and Killian & Teissier (1973) according to morphological and phyto-ecological characteristics, all cited by Konate (1996, p.15). These physical criteria determine their local use by societies. In turn, the resulting specificities determine the way in which they are managed.

As fertile zones, these environments are also characterised by their complexity in terms of physico-chemical functioning (Akassimadou et al., 2014, p.1) or ecology through morphology-soil-surface state relationships (Albergel et al., 1993, p.13). It is therefore a type of system in which different components interact with each other and are interdependent. As a result, as soon as one of the components fails, the system is affected and may suffer damage that is sometimes irreversible. This is illustrated by a study carried out on the land of Walo (Senegal), which reveals an upward degradation due to salinisation and the disappearance of the plant cover. Land that is not subject to seasonal flooding is particularly affected by the abrasive action of the winds, which remove the superficial part of the soil, which is also rich in mineral elements. This action, known as wind deflation, results in the degradation of the vegetation cover (Sall et al., 2011, p.17). In general, climate change is influencing the deregulation and transformation of the appearance of the lowland landscape.

2-2-2- Human responsibility for watershed erosion

Although much of the degradation of the landscape is due to human activity, responsibility for the erosion of catchment areas seems to be shared between natural causes and human factors. We would say that the actions are combined. Albergel et al (1992, p.128) note that in the Gagara basins (Burkina Faso), due to the disappearance of the shrub layer and especially the herbaceous layer as a result of the local agro-pastoral production system, there has been an extension of the heavily encrusted areas. This change in condition leads to an increase in runoff and erosion. This increase accentuates the effects of drought and leads to a very marked and sometimes irreversible degradation of the environment. We are therefore witnessing a trend towards increasing aridity of the environment, which can be described as the satellisation of the basins. This observation is confirmed by Perez et al (1995, p.62), who note that although the lowland soils are the most fertile in the region and are used for growing cereals (sorghum, rice) and off-season market gardening, the main agricultural constraints are due to the combined effects of the violent floods that sweep through the lowlands and the colluvial-alluvial inputs generated by the intense erosion of the slopes. Some of these floods are due to abundant rainfall over a short period of time. But a large proportion of them are linked to the degradation of the plant cover that forms a protective layer on the soil against erosion. Abusive and uncontrolled exploitation of fragile environments significantly reduces the protective function of vegetation.

Many researchers, including Bouzou et al (2009, p.152), point out that soil denudation continues as long as the traditional system of extensive cultivation without soil protection persists. This phenomenon is very prevalent in Sahelian environments in a context of very high anthropisation. The result is an increase in run-off on the slopes, leading to changes in soil surface conditions, gullying, widening and even deepening and silting up of watercourses. In fact, these Sahelian lowlands are located in a rather difficult environment with very fragile natural balances.

According to Bah et al (2017, p.4), the pressure on hillsides, which are constantly being degraded by bush fires, deforestation, soil impoverishment and mining activities, is causing silting and silting up. Almost all of these forms of degradation are caused by man's insatiable appetite for space and consumer goods taken directly from natural environments.

This situation is increasingly leading to lowlands being used for agricultural activities such as rice growing or market gardening. Moreover, the enthusiasm for lowlands in terms of the types of crops grown, the degree of intensification, and even the relevance of development, depends on the economic stakes represented by these sites (Lavigne et a1., 1996, p.10), which have become strategic for the agricultural development of certain countries.

Considering the ecological complexity of lowland environments on the one hand, and the benefits they provide to their users on the other, the ideal approach should be to take a realistic look at the issues involved in rational use.

For Jouve (1992, p.471), the interest of these lowlands is multiple: they allow diversification of production and provide work and income at a time that complements that of rainfed crops. Hence the need for professional management of these areas, taking into account their potential agrarian dynamics.

2-2-3- The need for lowland development

Kafilatou (2017, p.1618) notes that the use of inland valleys has become increasingly inevitable, as they occupy a key place in production systems (rice, maize, etc.) in the current context of climate change. Lowland farming can therefore be seen as an adaptation to climate variability. It compensates for the change in the cropping calendar that leads to lower yields and soil impoverishment. But it is clear that in Africa in general, socio-cultural and natural conditions are not always favourable to rainfed and irrigated rice growing. However, Africa has about 1.3 million km^2 of lowlands, and rice growing in these areas, using improved facilities, produces definite benefits that should be improved (Danvi, 1995, p.4). Since the constraints are enormous and the benefits essential for the population, the development of lowlands is therefore an obligation if the objective of effectively meeting food needs is to be achieved.
Lespez et al (2009, p.6) see the importance of the issue of food security in relation to a logic of land development and management leading to transformations in space and practices, and hence the landscape changes observed.
Landscape transformations or evolutions then introduce us to the general theme of adaptations. And in this perspective, Faho (1995, p.18) stipulates that the corollaries of the hazards linked to the qualitative and quantitative insufficiency of rainfall in the Sahelian region, the presence of sensitive slopes, and the continual degradation of natural resources under the effect of demographic pressure call for another type of natural resource management within the framework of agro-sylvo-pastoral development. This new problem is widely shared by most development structures and institutions.
A report by Benin's Ministry of Agriculture, Livestock and Fisheries states that, given that the country's growth potential depends largely on agriculture, which has a great deal of arable land that can be farmed, modernising the rural sector will enable the country to move from an economy dominated by self-subsistence to one that is much better integrated into the market, while at the same time creating jobs (2001, p.3). Given that local populations consume a great deal of rice and market garden produce, the many low-lying areas throughout the country must also be put to good use.
The social aspect of lowland development should not be overlooked. In this respect, Sidibe et al (2004, p.1), at the end of a study carried out in a number of localities in Mali, conclude that one of the rare opportunities for rural women to generate income, particularly in the cercles of Kadiolo and Bougouni, is to become involved in lowland rice cultivation. In fact, lowland rice accounts for 34% of the income of the women interviewed, and women who cultivate both

lowland and upland rice are relatively less poor than those who cultivate upland rice only. So the development of lowlands not only has economic benefits, it also helps to consolidate the family fabric and social stability.

Finally, from a sustainable development perspective, does the use of inland valleys contribute to supporting communities' fight against environmental degradation? Jouve (1992, p.471) emphasises the ecological advantage of developing lowlands, since this type of intervention has a particularly strategic dimension in the fight against aridity in the Sahel. Although the amount of water received by the soil per year seems to be insufficient to meet the needs of agricultural plants (700 to 850 mm), it is clear that beyond the poor distribution of rainfall, the particular ecology of the lowlands (their natural humidity based on the quality of the soils and their hydro-geomorpho-pedological characteristics) greatly minimises the impact of this poor rainfall. Hence the migration to fertile lowland areas. But to get the most out of these sites, which offer great socio-economic opportunities, it is necessary to manage the lowlands better. The aim is to protect them more effectively from degradation and make them more profitable with a view to sustainable development. This type of response is much better than reconverting farmers and other users to other activities that are certainly much more threatening to the environment because of the drastic drop in crop yields and repeated water stress (Bah et al., 2017, p.4).

However, lowland development requires appropriate means, techniques and policies based on a good knowledge of the interrelationships between the ecological system and the agricultural system, using remote sensing tools. This knowledge is therefore fundamental for understanding the evolution of environments over time (Devineau et al., (1991, p.377). A synergy of action is needed to ensure that the inadequacy or lack of a high level of technology among farmers, and the large reservoirs in the lowlands, do not cause more problems than they solve (Danvi, 1995, p.19).

2-2-4- Human adaptation strategies

The need to adapt social practices when using lowlands is explained by the constraints imposed by these sites. Successful adaptation theoretically offers a head start on the difficulties associated with occupation, a broad investment opportunity and maximises the benefits associated with exploiting the space. For Bah et al (2017, p.3), adaptation is an essential action to anticipate the impacts and thus reduce the costs and damage associated with the occurrence of climate change. But this is not always done in advance. As a result, the results do not always live up to expectations and sometimes lead to disaster. This is why training farmers, understanding their needs and regular technical monitoring are essential factors in the successful use of lowlands. With this in mind, Lespez et al (2009, p.5) point out that the typology of inland valleys reveals that there are facilities that seem appropriate for each type of inland valley and according to use. In other words, the development of lowlands is not a typical serial operation. Prior to the work of Lespez (op cit.), Zombre (1992, p.22) had illustrated this idea through his work in Yatenga (Burkina Faso). He says that the consequence of satelliteisation is an increase in morpho-dynamics: wind erosion and widespread intense run-off, prompting farmers to develop their own water and soil conservation techniques, such as the planting of stone cordons and the particularly promising Zai method. This is a particular method of cultivation in stacks, adapted to severely degraded pedological areas, which enables water and fertiliser to be concentrated in micro-basins where the seeds are sown. So, in a context where climatic characteristics can push a region towards Sahelisation, farmers' resilience is reflected in water retention and management techniques. Although these techniques have a number of points in common, the specific features of each region, local know-how and many other parameters give these practices a local configuration. Not all environments are the same. Since climatic disruptions to the ecosystem result in land erosion, flooding, the early drying up of ponds, production losses, lower agricultural yields, etc., the effects are diverse and felt differently. Consequently, to limit the effects of climate change and improve their well-being, local populations are developing adaptation strategies that vary according to their categories (Atidegla, 2017, p.1). Local adaptation strategies are developed according to whether there are natural or human constraints. With regard to these constraints, Lespez et al (op cit.) list two types: firstly, there are natural constraints caused by heavy flooding, submersion, insufficient rainfall and rapid weed growth on land cultivated in the lowlands. The second type of constraint is socio-economic and technical. In all cases, the response is limited to essentially technical solutions, generally oriented towards more or less complete control of water, i.e. irrigation.

However, water management takes account of local specificities. For example, in Côte d'Ivoire (Assouman et al., 2015), farmers are adapting to the alluvial plains of the Bandama River in two ways: on the one hand, they are moving plots from flood-prone areas to more secure parts of the land, and on the other, they are adopting new cropping methods and techniques that are either opposed to or associated with the current or empirical ones. In addition, proposals for an improved system based on intensive farming, the use of inputs and new varieties appear to be solutions for regulating the cropping calendar and responding to the attachment that farmers show to their land, Kafine in this case. Jouve (1992, p.471) presents another equally effective approach, which consists of investing in knowledge of the hydrological functioning of the lowlands to avoid serious problems during their development. In this approach, irrigation is an appropriate response to the fight against aridity, and makes it possible to meet the food needs of a rapidly growing population. Roose (1999, p.21) is more explicit in citing the traditional system of tanks as a system for managing erosion, making it possible to supply the lowlands in poorly permeable areas. These village reservoirs ensure that the hydrological deficit is compensated for by water flows in the surface parts of the soil and by fairly short concentration times. Lastly, Ngamine et al (1999, p.247) present the following

plan

a project to develop the Tchanar terroir (Chad) based on filtering bunds, which has enabled erosion to be kept relatively under control and crop yields to increase moderately (10 to 20%).

Farmers clearly cite erosion as one of the main constraints to lowland farming. But local differences in cropping techniques and environmental characteristics mean that farmers' adaptation strategies vary.

2-2-5- The contribution of the natural environment in central-northern Côte d'Ivoire

In West Africa, inland valleys represent a very important potential for agricultural production in general and rice production in particular. Rational use of this type of ecosystem requires an understanding of its pedological organisation and hydrological functioning (Sitapha, 1996, p.5). Côte d'Ivoire has opted to intensify agriculture in the plains and lowlands (Cheron et al., 1991, p.438) in order to feed its population from domestic production. To achieve this, synergy between all the regions is strongly recommended. These include the northern regions. North-central Côte d'Ivoire does not have the most abundant agricultural land. The soils emanating from the most important geomorphological landscapes generally belong to lateritic croups and fragile tabular plateaux. These landscapes have a low level of humanisation. The petrosteritic and gravolitic series occupies 2.6% of the area and is not very significant, but these soils have good water retention capacity, good drainage and very good cultivation properties (Koli Bi, 1999, p.119).

However, the centre-north has as much of an agricultural vocation as the other regions of the country. Brou (2005, p.42) considers that the lowlands of the savannah regions of the north have great potential for intensifying and diversifying agricultural production and are a place for conserving biodiversity. However, he goes on to warn that despite this potential, the use of lowlands is still limited, both in time (little use in the dry season) and space (many lowlands are still unexploited). Undoubtedly, the under-exploitation of lowlands in northern Côte d'Ivoire is due to a number of constraints, including land tenure problems, and no doubt difficulties in land management and tillage, etc. Diagne et al. Previously, Diagne et al (2004, p.3) had pointed to the slow growth in domestic rice production in Côte d'Ivoire, blaming it on low yields in rainfed upland and lowland rice production systems.

However, the lowlands are quite numerous and can make a huge contribution to Côte d'Ivoire's food policy, which is to some extent owned by women. Assi (2008, p.26) points out that women occupy a significant position in the production process and in the reorganisation of the rural economy in crisis. Their involvement in developing the lowlands has led to innovations in labour and food production. Developing the lowlands of north-central Côte d'Ivoire could open up new work areas for a dynamic rural population, but more than that, it could also reduce the vulnerability of these areas, which are subject to almost permanent flooding and cause the destruction of fields and other human structures (Assouman, 2015).

2-2-6- Characteristics of lowlands in Côte d'Ivoire

Over the last twenty years or so, demographic changes and the dynamics of land use have profoundly altered the potential associated with lowlands, whether or not they have been developed (Cecchi, 1992 p.274). These areas have taken on a new identity and are becoming increasingly important in the conquest of land for agricultural purposes. Agricultural intensification is opening the way to more intensive use of the lowlands for more comprehensive land use. But the lack of irrigation infrastructure makes it difficult to exploit these areas, despite their high potential. A study carried out by NEPAD in 2005 revealed that irrigated agriculture is of secondary importance to agriculture in Côte d'Ivoire, despite the significant potential offered by the lowlands and irrigable plains throughout the country. According to the same pan-African organisation, the total irrigated area in Côte d'Ivoire is around 32,500 ha out of a potentially irrigable area estimated at around 180,000 ha. This irrigated area is related to the construction of several dozen dams, including around twenty in the north, to irrigate low-lying areas that have been converted into rice-growing basins located downstream (Cecchi et al., 1992, p.19).

Today, in a context of climatic changes taking various forms, including water imbalances in some areas, there is a need for better management of hydromorphy in lowlands. According to Cecchi et al (op. cit., p.26), sustained and pronounced climatic pejoration, combined with changes in land use and surface conditions, is leading to significant alterations in the availability of water resources, in terms of both quantity and quality, in small West African catchments and their lowlands. Côte d'Ivoire is no exception. To cope with these unpredictabilities, farmers opt to build near water reservoirs that are generally perennial in normal rainfall conditions. These small local dams have real agricultural potential, but are still insufficiently exploited, and often conflict with the pastoral use to which they were initially put (Cecchi et al., 1992 p.26, idem). The economic benefits are therefore debated.

The identification of lowland farmers remains mixed. In some regions (Gontougo, Bandama valley, N'zi-Comoe, Lacs, Worodougou, Bafing, Savanes, Denguele), where lowland areas are of secondary importance in the rural economy, they are easily ceded to women, young people and foreigners, and the customary law land tenure system applied to them is less rigid (NEPAD, 2005 p.10).

On the other hand, in Bete country in the department of Gagnoa (Goh region), male head of household rice farmers accounted for 80% compared with 20% of female head of household rice farmers (Depieu et al., 2017 p.81). Chaleard (1988, p.46) made another classification of the actors based on socio-economic characteristics. In his view, two social groups dominate: landless peasants, who saw in the establishment of these lowlands the possibility of gaining access to an indispensable factor of production, and urban workers (industrial workers, shopkeepers, civil servants), for whom rice-growing ensures the subsistence of the family group at a lower cost and occasionally provides a marketable surplus.

However, these farmers do not use the lowlands exclusively for rice production. Local cultivation techniques are used to maximise the benefits of these areas. For example, as Chaleard again points out (op. cit., p.43), rice is generally grown in the rainy season, but some farmers dig holes in the lowlands to irrigate and obtain a more remunerative off-season crop (from January to June). Another technique involves varying speculations. According to the same author, lowland plots are cultivated for one year, combining rice and mats, and then left fallow. A few cassava plots have also been created on fallow land near the village or on old abandoned plantations. Lastly, market gardening is gaining in importance. These crops are increasingly valued for their rapid growth, large-scale consumption and relatively low cost on local markets. Tomatoes in particular, but also *ndrowa* (a local variety of aubergine), okra and chillies are taking the place of *poyo* bananas both in production systems and in space, since they occupy low-lying areas (Chaleard, 1988 p.43).

Despite all the competing speculations, rice continues to play a key role in lowland farming. However, rice self-sufficiency can only be achieved by improving the level of domestic production through the development of lowland rice cultivation. In fact, lowland rice cultivation in improved and/or unimproved mode represents around 36% of the total area under rice cultivation in Côte d'Ivoire and contributes to more than 20% of paddy production, according to Depieu et al. (2017) p.80). The same authors also report that yields are slightly lower in lowland rice cultivation without irrigation (1.53 t/ha) than in lowland rice cultivation with irrigation (1.70 t/ha). But an efficient policy for developing these areas without dams could boost production. Hence the need for inland valleys to be discriminated against, as the lack of inland valleys characterisation leads to standardised designs that are often not adapted to the specific conditions of each site to be developed (NEPAD, 2005, p.3).

2-2-7- The shallows of Katiola

On a national scale, the Katiola region could become a rice-growing centre, given that the geomorphology and pedological conditions are in many respects favourable to agriculture. In fact, most of this region (almost 60%) is covered by soils and models of average suitability for mechanised cultivation (Poss R., and Sabathe R ,1982, p.179).

The hydrological functioning and model of the lowlands depend on the local climate. In fact, depending on the year, the climate is similar to that of the more northerly zones, with a single rainy season, or to that of the meridional zones, characterised by the alternation of two dry seasons and two rainy seasons, which could also explain the relative anthropic pressure on these environments, as noted by Toure (1999, p.151). In fact, he explains that these marginal and previously untouched environments are now highly productive in both geographical and human terms. However, we note that contributions to the local economy are becoming a reality. In addition, these agricultural areas remain under-exploited due to a number of barriers of a social nature (an ageing and deficient labour force) and a physical nature (above all, problems with water management). It is therefore necessary to overcome these barriers to achieve optimum productivity in these lowlands.

2-3- DEFINITIONS OF CONCEPTS

2-3-1- Contextual meaning of the term "Problem

The term "problem" has a particular connotation depending on the environment and context in which it is used. It can be defined in many different ways, and each context gives the term its own explanation. In its most common sense, a problem is a situation in which an obstacle prevents you from progressing, moving forward or achieving what you set out to do. A problem arises when there is a difference between the way things are and the way they should be.

But the framework in which we use it in this work in geography gives it a meaning that goes far beyond the notion of obstacle. For this reason, we present it as a paradigm. In epistemology in the humanities and social sciences, a paradigm is a representation of the world, a way of seeing things, a coherent model of the world that rests on a definite foundation. This corresponds to the idea that, in our context of study, we need to present the problem as a constraining situation which obliges us as researchers to understand its origin, its manifestation and to envisage plans for adaptation or resolution.

In short, the problem is not a situation in which an obstacle prevents progress, but rather a situation or difficulty that needs to be resolved using a methodical and methodological approach.

2-3-2- Geomorphology

Geomorphology is defined as the study of the shape and revolution of the earth's relief. It is a branch of physical geography. Geomorphology is linked to geology (whose contribution is necessary to understand the lithological and tectonic structure), physics and chemistry, disciplines that help to explain the consequences of temperature variations, the physical actions of water and the characteristics of alterations.

2-3-3- Lowlands

Low-lying areas can be identified by topographical, morphological, pedological and hydrological criteria. Their soils are often waterlogged or submerged for a more or less long period of the year by a water table corresponding to outcrops of the water table and runoff. It should be noted that several research scientists have attempted to define a lowland. In relation to hydrology, Killlan and Teissier, for example, give the following definition of lowlands: "[...] the bottoms of small valleys, not very deep, regularly flooded in the rainy season by marshes with difficult and temporary drainage...". Of all the definitions that exist, the scientific community often agrees that Raunet's (1985, p.181) is the most complete and therefore the most interesting. Here is the expression:

"In the intertropical region, we call lowlands the flat or concave bottoms of valleys, small valleys and flooded drainage channels that form the basic drainage axes embedded in the thick alterations of the "peneplanised" crystalline basement. These are the preferred axes of convergence for surface water, hypodermic run-off and groundwater contained in the thick alteration mantle and fed by rainfall".

This is the definition we will use in this work, since it has the advantage of characterising lowlands in the general context of the landscape and in relation to morphological units.

2-4- RESEARCH PROBLEM

After reviewing the scientific literature relating to lowlands, we can conclude that a good number of publications have been made public by the scientific community. But very few of them deal with the geomorphological problems of these environments. Although some have tried, their analyses remain superficial and fragmented.

In Katiola, in *Tagbana* country, farmers generally encounter problems of water control, phytosanitary problems and productivity problems. However, geomorphological problems have rarely been addressed in terms of constraints on the use of lowlands. However, identifying and taking account of problems of geomorphological origin could make a considerable contribution to better management of these areas in order to optimise the yield of the crops grown there. While water management issues have been satisfactorily addressed in lowlands managed under dams, the situation is not as good in lowlands managed outside dams. In fact, the lack of adequate development in terms of water management infrastructure, site planning and sediment transfer reveals a lack of technical skills on the part of the farming population in managing the problems associated with the dynamics of the lowland. Moreover, the relationship between site geomorphology and crop productivity is not always highlighted in studies of lowlands. However, lowlands are sites in a geomorphological context with characteristics that may contain micro-model imperfections, which expose them to water control problems and, as a result, to a major agricultural production deficit. Hence the importance of controlling their dynamics, i.e. the effects of erosion and site morphology.

We therefore need to look more closely at the inherent characteristics of the lowlands developed outside the dam in order to define run-of-river profiles capable of responding effectively to farmers' expectations.

The current initiative to study these problems on the scale of a lowland in the locality of Katiola is a pioneering one. It aims to make more explicit the problems of erosion in lowlands that are farmed outside a dam, and to identify the resilience mechanisms put in place by the farming community. Efficient use of the lowlands developed outside the Katiola dam, in particular the lowland that is the subject of this study, therefore requires a detailed descriptive examination in order to identify the erosion problems that are likely to arise, with a view to better defining the adaptation strategies developed locally by farmers.

So the main question is: **faced with geomorphological constraints, what solutions have farmers found**?

The general aim of our study is therefore to find out how farmers are resilient in the face of geomorphological constraints in the lowlands outside the dam at Katiola.

Specifically, this involves :

- Describe the type of non-dam lowland development studied;
- Identify the geomorphological constraints encountered in the low-lying area under study;
- Identify resilient solutions or responses to these geomorphological constraints.

3- QUESTIONS AND OBJECTIVES

3-1- REMINDER OF THE RESEARCH QUESTION

What solutions have farmers found to geomorphological constraints? An in-depth study of this question requires us to analyse it in the following way:

- What physical indicators explain the presence of problems of geomorphological origin?
- How do geomorphological problems manifest themselves in the study area?
- What solutions have local operators put in place to deal with problems of geomorphological origin?

3-2- MAIN AND SPECIFIC OBJECTIVES

To gain an understanding of the ways in which farmers are resilient to the geomorphological constraints of the lowlands developed outside the dam at Katiola. Specifically, the study aims to :

- Describe the type of non-dam lowland development studied;
- Identify the geomorphological constraints encountered in the low-lying area under study;
- Identify the solutions developed by farmers in response to these geomorphological constraints.

II PART

MATERIALS AND METHODS WORK

Part Two:

MATERIALS AND WORKING METHODS

This section sets out the scientific mechanism used to carry out this study. It highlights the collection method and materials used to achieve each objective.

4- RESEARCH METHODOLOGY

4-1- ELEMENTS AND METHOD OF WORK 1

4-1-1- Reminder of question 1

What are the physical indicators explaining the presence of problems of geomorphological origin?

4-1-2- Reminder of objective 1

Present the geomorphological environment of the low-lying area outside the dam being studied;

4-1-3- Data acquisition mode 1

Documentary research: our research led us to explore theses, articles, dissertations and many other scientific publications. It was carried out in libraries, in particular that of the Institut de Geographie Tropicale (IGT), the library of the Institut de Recherche et de Developpement (IRD) and the national library. The aim of the operation was to bring together existing information on the characteristics of the physical environment of the lowlands in the humid tropics in general, and in the Katiola region in particular.

Qualitative research: this aims to describe the lowland through its properties and the relationships between its components. This research consists of identifying the components of the environment that enable it to be characterised and defined, particularly in terms of how it functions. More specifically, it involves eliciting and collecting data in the field. This type of research provides a better understanding of how elements of the local environment play a role in the occurrence of geomorphological problems in the lowlands.

4-1-4- Data type 1

Satellite data: The image we have chosen is a panchromatic image from the Shuttle Radar Topography Mission (SRTM-1). The slab selected covers the square degree of the study area. It was acquired on 3 October 2019 at www.geonetwork.icrisat.org. This image has a resolution of 30 m and is georeferenced in the UTM-WGS84 coordinate system. The main advantage of SRTM data lies in the nature of the signal used, i.e. a radar signal. This is independent of cloud cover. This reduces or even eliminates the risk of errors due to the presence of clouds. This is a major advantage for inter-tropical regions such as the Ivory Coast, where cloud cover is almost permanent. What's more, the images are easily accessible via online platforms. The image was processed using

GIS software to generate a DTM of the square degree in which the studied lowland is located.

Agrarian morphological data: this involves a physical description of the low-lying area with the run-of-river profile studied. Using photographs and field surveys, we determined the area of the lowland, the structure of the perimeter and the hydraulic operating system.

4-1-5- Collection equipment and tools 1

We used technical equipment to collect and process the data. These are the following instruments:

For collection,

- A GPS receiver is a geolocation system that uses a satellite signal receiver to pinpoint the coordinates of points of interest;
- A digital camera to take pictures of some rather special sites to support and illustrate what we're saying;
- A decameter for measuring length;

For treatment,

- Mapping software for operations relating to geographic information systems in order to facilitate spatial and thematic processing;

4-1-6- Variables 1

We were interested in territorial variables.

- Land morphology variables (surface area, structure of the perimeter, number of basins)
- Morpho-dynamic or topographical variables (slope measurements, landforms)

4-1-7- Collection techniques 1

The data required to achieve our objective were collected in two main stages on sections of the lowland studied, which constituted our sampling unit. The first involved obtaining an aerial view using remote sensing. We then collected additional data on the site using field collection techniques. We are particularly interested in a specific part of the lowland rice-growing area. This portion is defined as the observation sequence. It was selected because of the presence of hydraulic trunks, the representativeness of soil units and the presence of alluvial sections. Observation plots of approximate size were selected to represent the observation units. The aim of this procedure is to identify illustrative and explanatory elements of geomorphological problems. This is the purpose of the present study.

4-1-8- Analysis method 1

The aim is to describe the lowland on the basis of its intrinsic properties and the geomorphological context in which it is located. The data enable us to identify the elements required for this purpose. We cross-referenced two or more factors in order to understand how these relationships determine the nature of the lowlands. Using images, we tried to design a representation of the relief using parameters such as contour values. Ultimately, we produced a digital terrain model. The DTM produced is useful because, even though the spatial resolution of the image does not allow us to clearly identify the lowland under study, it does allow us to perceive, from the relief, that the area of interest is likely to contain a large number of non-dam lowlands. Furthermore, it is thought that the conclusions of a study carried out on one of these non-dammed lowlands could be applied to the others.

4-2- ELEMENTS AND METHOD OF WORK 2

4-2-1- Reminder of question 2

How do geomorphological problems manifest themselves in the study area?

4-2-2- Reminder of objective 2

Identify the geomorphological constraints encountered in the lowland studied.

4-2-3- Search type 2

Literature search: the aim of this literature search is to bring together existing information on disturbances that may occur in lowland areas. We focused on disturbances related to soil texture, sediment flux or the peaty character of the lowland.

Qualitative research: this aims to assess the properties of the lowland as it is and the impact it suffers from the environment in which it is located. This research aims to understand how lowlands are affected by erosion. We then show the causal relationship between the site's micro-patterns and the geomorphological dynamics of the lowland. The basic concern is to find out whether the population is really aware of the existence of problems in the lowland linked to the characteristics of the relief, before finally realising that there are problems of geomorphological origin. Here, we are dealing with the *perception of* problems of this type in the lowlands they use.

Exploratory research: We have a small amount of data on the fagon whose low-lying areas managed outside the Katiola dam, and more specifically the one that is the subject of this study, are exposed to geomorphological problems. Consequently, we have carried out field observations in order to produce the data we need.

4-2-4- Data type 2

- Geographical data: using topographical maps, we determined how the setting already exposes the lowlands to geomorphological risks. To gain a better understanding of these risks, we carried out a more detailed, large-scale analysis of a segment of the lowland.
- Experimental data: this is data that we were looking for in the field. They

were acquired through visual observation and tactile analysis. They include traces of erosion, exondation and flooding.

- Empirical data: these are the comments we collected from lowland farmers about the difficulties of developing the lowland.

4-2-5- Collection equipment and tools 2

- The interview guide: based on a multi-theme structured plan, we talked to a sample of fifteen people using the lowland in order to obtain information.
- Technical equipment: this includes the essential instruments for collecting data. These include dictaphones, meters, filters, sieves, etc.
- Manual work equipment: this is a set of manual tools used to facilitate exploration of the site (essentially a machete).

4-2-6- Variables 2

We were interested in technical variables.

- Static or topographical variables (shape of the bottom of the lowland, general appearance of the lowland, presence of outcrops, topographical facet, etc.).
- erosive dynamics variables (level of bank degradation, alluvial or erosion constraints, alluvial characteristics, biological activities, etc.).
- Geo-pedological variables (size, sediment density, soil permeability, termites, etc.)

4-2-7- Collection techniques 2

- The interview: with the help of a guide, we tried to find out from a sample of the population farming in the lowlands what difficulties they encountered in their farming activities in relation to the terrain. The sample was made up of leaders or representatives of the farmers. They therefore voluntarily answered the questions under the supervision of their peers.
- Experimentation: more specifically, this involves provoking and collecting information in the field from physiographic units. These physiographic units are portions of land with a similar edaphic character, in particular the same soil or combination of soils, and similar morphodynamics under similar hydrological constraints. Such a physiographic unit is assumed to react to natural agents and human activity: in other words, the same conditions apply almost homogeneously everywhere in the same lowland. Each physiographic unit can then be used to observe one or more phenomena. This form of research has enabled us to gain a better understanding of how elements internal to the local environment play a role in the occurrence of geomorphological problems in a lowland. With this in mind, we carried out tests and took samples in order to understand the dynamics of the environment and its effects on the structure of a lowland.

4-2-8- Analysis method 2

By interviewing the local people and obtaining data on the dynamics of the lowland, it was possible to relate the constraints they expressed to the physical environment in order to establish the presence of geomorphological problems in the diagnostic environments.

In addition, the extraction of a segment of the lowland over which a large-scale analysis had been carried out provided a physical support compatible with man-made layers. This method made it possible to see in detail the topographical organisation of the relief, the occupation and use of the land and to measure the vulnerability of the issues at stake if the hazards were not controlled.

4-3- ELEMENTS AND WORKING METHOD 3

4-3-1- Reminder of question 3

How is the farming community in the low-lying, non-dam floodplain demonstrating its resilience?

4-3-2- Reminder of objective 3

Identify the solutions developed by farmers to cope with geomorphological constraints.

4-3-3- Type of search 3

Qualitative research: the aim of this research is to find out how farmers cope with the erosion-related constraints affecting their working area. We are looking here at the adaptation techniques that these farmers put in place in order to resist as much as possible the constraints imposed by the natural environment.

Quantitative research: this involves quantifying or estimating data with quantifiable properties in the field. Using a suitable data collection tool, values such as the area of cropland abandoned or added as a result of geomorphological constraints, the amount of production gained or lost, the number of farmers relocating or leaving the lowland, etc. were recorded.

Exploratory research: this consisted of discovering the adaptation techniques implemented by the farmers by accompanying them and observing them closely as they carried out their tasks. Where necessary, we used discreet and non-threatening questions to refine our understanding. For example: "What do you do in the event of flooding or the risk of flooding to protect your crops?

4-3-4- Data type 3

- Factual data: this is data collected from observations in the field (nature of the material, strength of the material, type of crops, cultivated areas). They have been acquired through observation and experimentation.
- Empirical data: these are anecdotes and facts gathered from lowland farmers through verbal exchanges. Their aim was to tell us about their know-how in managing the movement of water in the area they farm.

- Aerial photography: a *Google Earth* aerial photograph from August 2019 relating to the lowland analysis sequence was obtained in order to get an overall view of the site. The projection system, georeferencing and colour level were of acceptable quality, enabling us to extract the information we were looking for. Using GIS software, we applied processing to the photograph to optimise the classification and achieve a better characterisation of the land cover.

4-3-5- Collection equipment and tools 3

We used three types of tool to collect the information needed to verify this hypothesis.

- The questionnaire: using specific questions included in the section on objective 3, we recorded the techniques used by the farmers.
- The observation grid: this enables us to see how farmers go about their activity in a concrete way. By obtaining factual data, the observation grid makes it possible to fill in the gaps in the verbal information recorded and to go further than what was reported in the questionnaire.

4-3-6- Variable 3

We will be looking at socio-economic variables.

- Technical reinforcement: relating to the revaluation of the surface state (evolution of the surface areas, improvement of the plots, quality of the planting, type of agricultural equipment)
- Chemical reinforcement (use of fertilisers, selection of new seed varieties, frequency of plant protection treatments)
- Professional reinforcement (agricultural training, membership of a management structure, external assistance, pooling of activities, level of investment)
- Strengthening agriculture (diversification of crops, abandonment of the lowlands, cultivation of plateaux, combination of agriculture and fish farming)

4-3-7- Collection techniques 3

- Survey: this enabled the representatives of those involved in lowland farming to have their say. These discussions were based on individual and structural interviews.
- Observation: we used the active observation method. This is a process that consists of going as far as immersion in order to gain a better understanding of the workers' realities and see how the implementation of certain strategies constitutes a bulwark against local problems.

4-3-8- Analysis method 3

A content analysis is carried out on the data collected. This method of processing has enabled us to identify specific trends and establish relationships between them. For example, we analyse whether the type of soil in a portion of the lowland influences the type of crop grown by the farmer. This involves describing and analysing land use. The aim is to see and interpret the distribution of activities within the study area. Using aerial photographs, we proceeded to characterise the objects identified and their representations. This is a human method of photo-interpretation, based on the experience we have gained by going out into the field. Photo-interpretation is defined as the interpretation of aerial photographs and space images. The aim is to analyse complex phenomena at different working scales on the basis of their shape and level of luminescence. Photo-interpretation is a logical and reasoned approach based on the visual examination of images by an operator called a photo-interpreter. This logical approach is based on object recognition using reasoning, whereby the operator identifies homologous areas based on colour, texture and structure. The identification of geographical units therefore consists of characterising homogeneous areas, examining their surroundings and looking for similar areas.

III PART

RESULTS AND DISCUSSION

Part Three:

RESULTS AND DISCUSSION

This section presents the results obtained from this study and opens a perspective for reflection
by situating the present work in relation to previous studies.

CHAPTER 1: EDAPHIC CONTEXT OF THE LOWLAND AND GEOMORPHOLOGICAL RISKS

Lowlands are complex environments whose functioning is governed by local water conditions and the geomorphological configuration of the interfluves that surround them. Although considered to be fertile areas, these sites are not fully utilised in West Africa, particularly in Côte d'Ivoire, due to the risks they present in non-dam situations.

A general presentation of the environment will provide a better understanding of the phenomena that take place there and the resulting problems.

5- PRESENTATION OF LOWLAND PARAMETERS

The choice of the lowland which is the subject of the present study is guided by the desire to understand what is happening in a lowland which has been anthropised, but which has not yet been the subject of any dam-type development and the associated irrigation infrastructures. The Katiola lowland has a run-of-river irrigation system, a more or less informal form of irrigation that is fairly common in the region. In this case too, it is possible to observe resilient techniques on the part of the farming populations. This lowland was chosen, on the one hand, for its representative character on the local scale of lowlands developed outside dams, in a context with specific geomorphological characteristics (topographical profile, type of relief, micro-model of the lowland, etc.). On the other hand, because of its proximity and easy access (it is located on the Katiola-Dabakala road axis, just after the Katiola municipal cemetery), it can easily be monitored.

5-1- PRESENTATION OF REGIONAL PARAMETERS

The watershed in which the lowland is located lies to the east of the town of Katiola. The area lies between 8° and 9° north latitude and between 5° and 6° west longitude. The department has no catchment area of its own. According to Coulibaly (2009, p.88), it straddles the Bandama and N'Zi river basins. It is crossed in the centre by the Katiola-Dabakala road, which runs west-east through the Timbe and Bonieredougou sub-prefectures.

The topography is relatively flat on the southern fringe, rising gradually towards the north. The relief pattern helps to speed up the flow of watercourses. In the northern part, run-off is more pronounced because of the steep slopes. In the south, the relatively flat topography helps to slow run-off and contributes to the appropriation of depressed areas for agricultural purposes. These regions are rich

in low-lying areas. It is
In fact, this is where the one that is the subject of this study is located (Figure 2).

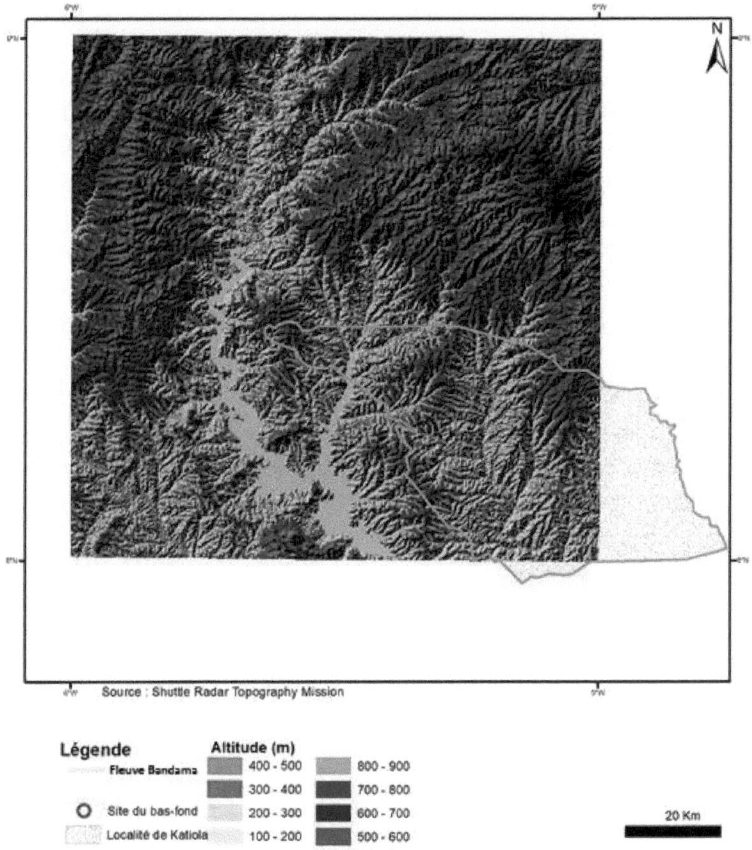

Figure 2: Relief of the square degree locating the lowland

Realisation: Bini E.

5-1-1. An uneven terrain

The general appearance of the lowland is not very irregular. The elevations are relatively average in relation to the general altitude of the locality. One of the highest points in the lowland rises to just 300 m, while the average altitude of the area is slightly above 400 metres, with a drop towards the south-east in the Katiola area, where the altitude is around 250 metres. The slopes are generally gentle, hovering around 5%. They are almost never sinuous and are only slightly steep. In places, the edges have steep slopes of around 15 metres, which drop directly into the shallows. The slopes are neither folded nor cut, but are subject to the impact of the wind, which changes their appearance, turning them into windfaces

that are partly decapitated and partly herbaceous.
The basin covered by this study has a general NW-SE orientation and covers an area of around 3 km^2 , characterised by rice-growing plots and as yet uncultivated fallow land.
The upstream part is marked by basic, more fertile soils with a finer texture. These areas are characterised by the presence of silty and clay-loam soils. Downstream, the model changes to slightly more accentuated depressions with sedimentary formations made up of eolian sediments, alluvium and, in places, lateritic crust.

5-1-2. Main morphological features

The vast outcropping tabular plateaux, such as those found at Katiola, are one of the major morphological features of the regional landscape. Induration appears to be an important process, and is very common in this part of northern Côte d'Ivoire. The landforms here are essentially remnants of laterite armour resulting from the leaching of loose elements (sand, gravel, etc.). From this erosion process emerge armoured interfluves cut by the sinking of a hydrographic network, especially according to the axes of water circulation, such as the fractures in the crystalline basement. From the lower centre, towards the south-east, there are alternating plateaux and depressed areas.

5-1-3. The river system

The basin is almost equidistant from the Bandama river to the west and the N'Zi river to the east. In practical terms, it is located about 30 km either side of the beds of these two rivers. As a result, it benefits from a water supply characterised by the presence of several branches describing winding itineraries, with rapids in places. After numerous meanders in a north-south direction, the lowland is irrigated at a rate dictated by the topographical profile of the catchment. Characterised by medium to gentle slopes, the flow is relatively average, especially during periods of normal rainfall. The hydrographic network is divided into main drainage channels, secondary channels and tertiary channels. The main drain in the talweg provides sufficient support for the volume of run-off during average and low-water periods. The somewhat flat aspect of this area means that the water spreads out in the shallows. However, during rapids, with a current that is sometimes very strong and irregular, the hydrographic network is virtually non-existent. This network gives way to a tributary or, at best, an estuary.
Because of its grassy savannah and shrub savannah vegetation, this region is highly susceptible to run-off and armouring, so its relief is marked by alternating glacis and depressed areas.

5-1-4. Longitudinal profile

The longitudinal profile of the catchment is relatively regular and not very steep (Coulibaly, 2009, p.39). This is partly due to the general shape of the landform, but also to its human development. Sloping towards the south-east, it includes some medium-sized rocky or organic knolls in places. The slopes are slightly steep, but steep on the sides.

The longitudinal profile alternates horizontal and sloping reaches in order to regulate the passage and volume of water needed to irrigate a rice-growing plot. The profiles around the perimeter of the basin are characterised by convex profiles due to the fact that the gallery forest is present almost everywhere in the upstream basin. In places, patches of this convex land are used for agricultural purposes.

Within the basin, disturbance is low in the downstream section due to the almost flat topography. The undermining of the banks upstream and the run-off of decayed sediments washes into the downstream part of the basin, creating almost permanent silting.

The hydrological regime and the structure of the river system in the basin depend largely on the topography and climate, but also on the nature of the various geological formations crossed. In the northern zone, for example, the hydrological regime is more pronounced because of the proximity of the watercourses that feed the basin and because of the predominantly hard, clayey texture, which favours lateral flow over infiltration. In the southern zone, the hydrographic network is more uneven and stagnant or infiltrating depending on the season, due to the predominantly sandy texture.

5-1-5. The bedrock of the catchment area

The watershed, the spatial framework of the study, is entirely located in heterogeneous clay and sand formations with ferruginous armour outcropping or sub-outcropping at the top of the interfluves, while silty sands line the beds and talwegs of the lowlands. These sedimentary formations, which show levels of ferruginous induration resulting from the degradation and crumbling of the dismantled armour components reflecting a reworking process, are marked by the presence of an almost permanent water table. The water table is widespread in the heterogeneous clayey sand and sandstone. It acts as a support during periods of low water and is fed by floods during the peak rainy seasons.

Sedimentation is mainly detrital (sand, sandstone) upstream and finer (clay) downstream. This distribution is due to the fact that flow carries away the lighter materials and tends to leave the heavier ones in place or a short distance away. The formations that make it up follow the traces of cycles. Over the course of the seasons and with the rhythm of run-off, the erosion process has favoured a natural sorting according to the size and weight of the sediments. Thus, in the spatial distribution of materials, coarse elements have gradually migrated towards the south of the region. However, even in the upstream part of the region, there are pockets of coarse material, many of which are the result of disaggregation of the armoured slopes. The current forms of occupation of the lowlands are closely linked to the current soils, which are largely hydromorphic in the talweg and ferruginous on the slopes.

5-2- SEQUENTIAL PRESENTATION OF THE LOCAL MORPHO-PEDOLOGICAL CONTEXT

5-2-1. Terrain

The relief is almost uniform in the lowlands and not very steep. It is the local topographical variant of a lowland site included in the North Ivorian peneplain, with average local altitudes of up to 300m. The lowland is a slight depression with a depth of less than 30 m (*Figure 3*).

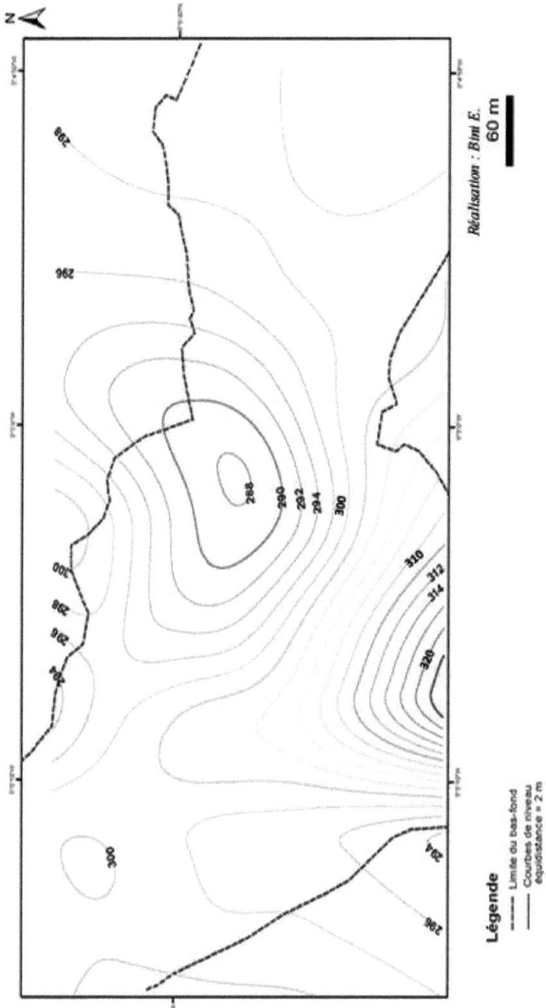

Figure 3: Topography of the lowland analysis sequence

Source: Google Earth, 2019

5-2-2. Generally gravelly interfluves

Residual armoured mounds, armoured and/or gravelly plateaux form the dominant landforms around the lowlands in the Katiola sector. These low-lying areas are the main forms of dissection in the peneplain. The more or less armoured slope soils or glacis are developed in the north on more or less clayey gres and in the south on more diverse formations. The lowland is settled in heterogeneous sand and gravel. Pedological samples from the main sites reveal Regosols on gravelly material and soils that have evolved little as a result of hydromorphic inputs. Lithosols on armourstone are confined to slope breaks where the armourstone outcrops in convex form. They appear sporadically over the lowland area in places where the geological unit of the substratum corresponds to this type of formation. There also appear to be discontinuous points of more or less high outcrops, distributed according to the extent of the relief and the nature of the bedrock. Their convex shapes make them difficult to penetrate. The relatively gentle slopes are often essentially gravelly, with eutrophic brown soils as alteration products.

5-2-3. Pedology of spreading glacis

From the high points defined by the plateaux or hillocks in places, the slopes taper off towards the drainage axis. This type of topographical profile contains two soil units: on the one hand, the lithosols on dismantled armour found in the upstream section, with a texture containing a high proportion of gravelly material. These formations correspond to ferruginous leached soils resulting from colluvium along the left bank; on the other hand, downstream, there are hydromorphic input regosols on gravelly colluvial-alluvial material. The gravelly granite texture dominates the lower slopes.

5-2-4. Lowland soils

In the broadest sense, the lowland soil is the pedological sequence occupying the valley bottoms: the morpho-pedology here integrates colluvial-alluvial micro-terraces with valleys. But on a larger scale, i.e. very close up, the details of their physiographic and pedological characteristics become clearer.

5-2-4.1. Hydromorphic soils on gravelly material

This unit covers the drainage axes and the northern part of the lowland. The soil has a cambic texture, showing signs of alteration to a greater or lesser extent at sub-surface level. Their structure varies from clay to brown sandy-clay, with a fairly gravelly surface (*Photographic Plate 1*). The organic matter content is fairly low and, as a result, so is the fertility. This is the hallmark of very filtering lowland soils. The excessive permeability of these lowland soils leads to significant mineral leaching.

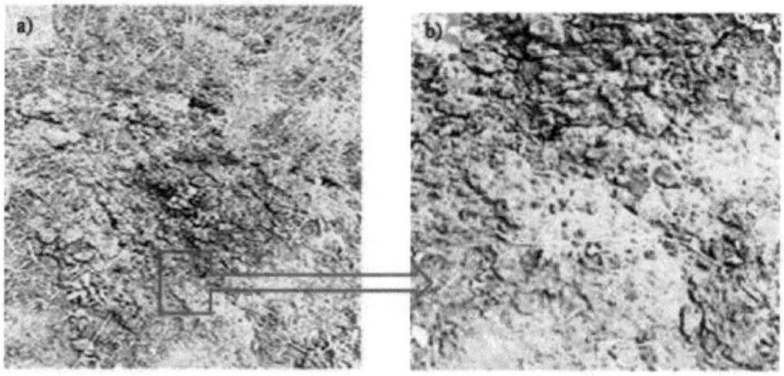

Photographic plate 1: Cambic soil composed of aggregates of clayey or sandy-clayey materials in a sub-surface horizon.

Cliche: BINIE, 2019

5-2-4.2. Hydromorphic soils on alluvial material

This unit mainly occupies the lowland in its southern part. It is mainly made up of hydromorphic soils on the surface and overall on alluvial material that varies but is often silty or silty-clayey. These soils, known as melanic because of their dark grey or black colour, are in fact rich in organic matter (*Photographic Plate 2)*. They have a fairly homogeneous and crumbly structure. Chemical fertility is highly variable. It is generally deficient compared with organic fertility.

Photograph 2: Melanic soil with very humified organic matter and traces of erosion

Cliche: BINIE, 2019

5-2-4.3. Ferruginous soils with concretions and stains

They are developed on colluvial-alluvial material. Most often, these soils have a clayey texture that is impermeable on the surface (*photographic plate 3*). They are grey or brown in colour. These soils generally show traces of gley on the surface. This type of soil, known as anthracite, is scattered throughout the lowlands. This type of soil has a low organic matter content and relative fertility.

Photographic plate 3: Anthraquin soil susceptible to crumbling and erosion in aerobic conditions

Cliche : BINI E., 2019

6- PHYSICAL PROPERTIES AND GEOMORPHOLOGICAL RISKS

Topography is a dominant factor in geomorphological hazards. It plays a key role in the erosion of slopes and the increase in sediment load in basins and therefore in lowlands. In addition, climatic, edaphic and geological interferences have an impact on the level of these risks, both in their evolution and in their geographical distribution within the lowland. The lowland is highly dependent on the topographical setting in which it is embedded. For example, it is possible to observe sedimentary formations that include levels of indurated materials of the fragisterite or petrosterite type. These formations can be explained by the dismantling of breastworks upstream. The resulting run-off develops the progression of the alluvium to varying extents and consequences in certain areas of the lowland. Even if wind action cannot be ruled out, the geomorphological risks to which this lowland is exposed are essentially due to the action of wintering.

6-1- RISKS OF ABRASION ON SLOPES

The low-lying area under study is embedded in a basin developed outside the dam. As a result, the sandy-clay soils identified in places contain material from slopes of varying size and relatively steep gradients. The region's climate is characterised by an alternating rainy season from the end of May to the end of October and a rigorous dry season. Depending on the year, the climate is identical to that of the northern zones, which have a single rainy season, or to that of the southern zones, which have two rainy seasons alternating with two dry seasons. In the latter case, the rains are generally spread over six months, from April to June and August to October. These rains have an erosive effect on certain parts of the lowland wall. This effect is marked by heavy run-off on rough soils caused by the length of the dry season, the roughness of the substrate or the lack of plant cover.

Near the downstream end of these slopes, it is possible to observe sterite and other materials transported by runoff down the slopes (*Photo 4*). Some of the depressions, generally those located close to or in the current of these slopes, contain ferruginous debris. Some water channels are not spared. In places, coarse sand and alluvial gravel are found, particularly towards the western boundary. Soils on gravelly material predominate in the western sector of the lowland. They are scattered from the cuirass scree accumulated at the foot of the escarpments (steritic megaepilite). These are therefore soils derived from successive abrasive cycles that affect the slopes. Channels are often affected by the phenomenon of obstruction due to the presence of material in the hydrographic network linked to the almost flat topography of the lowland and the proximity of desaggregable units (*Photographic Plate 5*).

Photographic plate 4: Plinth soil marked by a high concentration of clear nodules resulting from the alteration of ferruginous slopes (plinth or nodular material coated with iron oxide).

Photographic plate 5: Presence in the channels of net colluvial nodules resulting from the alteration of ferruginous slabs on the slopes.

Cliche : BINI E., 2019

6-2- RISK OF FLOODING IN THE LOWLANDS

As well as connecting the lowland to regional watercourses via tributaries, water is also supplied either by rain or by run-off from the slopes after a heavy rainfall. The latter mode of supply differs depending on whether or not there are facilities for this purpose. Drainage from the slopes can be channelled through small structures or, in the majority of cases, it takes place by surface run-off from the slopes.

But the presence of these natural microstructures for collecting and diffusing water in the lowlands is almost non-existent. In addition, in order to design any small structures in the lowlands, it is first necessary to estimate the amount of water flowing at different timescales (daily, monthly and annually). These data are all the more important as climatic hazards are increasingly exposing the lowlands to rainfall regimes for which farmers are not prepared. This state of affairs could be explained by the fact that a development is only beneficial if it is economically viable.

profitable. A development with total water control (dam and irrigation) would not be profitable for the areas currently under cultivation.

The small sewage works in the lowlands have difficulty withstanding the high pressures and large quantities of water. The fact that the levelling is not entirely regular accentuates the concentration of water in certain areas of the lowland. As a result, the centre of the lowland is more waterlogged. The volume of water gradually decreases towards the walls (*Figure 4*).

6-2-1. Run-off

The Katiola lowland has a rainfall of around 1,000 mm. It is characterised by medium slopes and relatively infiltrable soils. Runoff is therefore constant during normal periods.

But in very wet periods, the high rainfall coupled with the nature of the geological substrate and the morphometry of the basin will generate relatively high runoff in the lowland. This sometimes results in the plants being washed away by the current or being completely covered over for a few days. The phenomenon of rice lodging during flooding is quite common here. In fact, the variety of rice grown here does not have the same flotation properties as the long-stem rice grown in the deltas of major rivers such as the Ganges or the Brahmaputra, or even the Mekong (FAO, cf. fao.org). The risk of flooding is relatively high during the months when rainfall peaks (May-June and August-September). They are also increased at the beginning and end of the rainy season because of the degraded vegetation cover and the saturated hydrographic network. Runoff is therefore sensitive to soil and plant cover conditions. It should be noted that the last two years have been very rainy in Katiola, according to eyewitness accounts.

6-2-2. Infiltration

Like run-off, infiltration depends on the nature of the soil and, above all, its surface condition. Of course, low and continuous rainfall will play an important role in groundwater recharge: this is the soaking rain identified using a rain simulator during laboratory studies (cf. E Roose, 1977). Infiltration makes it possible to recharge underground reserves in order to meet the water requirements of plants during dry periods. However, as the lowland is essentially located in a depressional region, it is generally irrigated by the last passages of water that are still active. With a predominantly clayey texture, water retention in the upper layers is effective over a fairly long period. As a result, the central part of the lowland is not as dry as in the Sahelian region, as the water table is not very far away. In addition, during the rainy season, the slope effects associated with the nature of the slopes determine the violence of the floods in the lowlands. The soil in the lowlands is generally permeable at the start of the rainy season, facilitating infiltration, but once saturated, it becomes impermeable. Later, the soil becomes compacted.

6-2-3. Evaporation

During periods of flooding, evaporation predominates in the lowlands. The predominantly clay soil slows down infiltration and the thin foliage does not constitute a major obstacle to the infiltration of solar rays. However, this is not the case on the slopes, where evaporation is essentially the result of plant transpiration: this is known as evapotranspiration. Vegetation in these areas is denser.

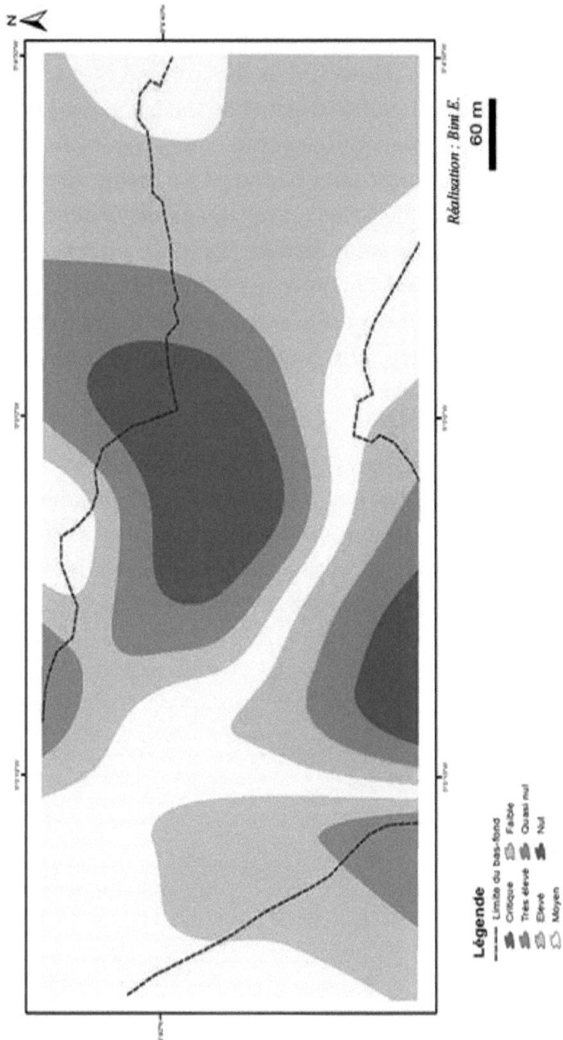

Figure 4: Flood vulnerability level of the lowland analysis sequence

Source: Google Earth.

CHAPTER 2: LOCAL IMPLEMENTATION OF PROBLEMS OF GEOMORPHOLOGICAL ORIGIN

The problems of geomorphological origin in the Katiola-Gnomonkaha lowland are linked to the structure and general dynamics of the surrounding slopes, to which are added the effects of the surface hydrodynamics of the lowland in its present state. Sometimes difficult to perceive before the area was appropriated by the farmers who use it, these problems seem to be omnipresent. To the point of becoming real dirimant factors in the economic productivity of this lowland, which is regularly developed in the eastern suburbs of the town of Katiola. These are obstacles imposed by current external geodynamics, the expression and continuation of sub-current morpho-climatic cycles, in line with the morphogenic explanations proposed by Filleron (1995) for the dynamics of the slopes of northern Côte d'Ivoire.

7- MARKS AND IMPACTS OF MORPHO-DYNAMICS

The geomorphological setting plays a key role in the fluidity of water in the lowlands. Depending on the structure of the relief, the variety of materials and their spatio-temporal evolution, the behaviour of this setting can be hostile or favourable to good development of the lowland. Thus, the development of the Katiola-Gnomonkaha lowland is closely linked to the environmental characteristics of the surrounding slopes and to the internal dynamics of the lowland (Table 1). This is even more important because the main lowland crop, rice, is highly sensitive to fluctuations in climatic conditions and in the flow rate of the water table during the rainy season. Thus, good control of the water level, taking into account geomorphological parameters (slopes) and potential hydrodynamics, could encourage good coordination between crops and water availability. This could also be a way of combating local soil impoverishment and degradation.

Unfortunately, the level of response, as will be shown below, reflects either powerlessness or an underestimation of the natural constraints present.

From this point of view, we will use the multi-criteria analysis framework proposed by Albergel (1993), which is illustrated in a synoptic table presenting different situations.

Parameters to be studied for lowland development	Regional criteria				Local criteria			
	Climate zones	Geology and relief	Demographics	Integration into the market economy	Basin size and morphology	Social and land organisation	Farmers' water management skills	Access to sources of finance
Hydrological functioning	++	++	0	0	++	0	+	0
Hydrogeological functioning	++	++	+	0	+	0	+	0
Floors	++	++	+	0	+	+	+	0
Vegetation	++	+	+	+	+	+	0	+
Lowland cropping system	++	+	+	++	+	+	++	++
Place of the lowland / production system	++	+	+	++	++	++	++	++
Possibility of intensification	++	0	++	++	+	++	++	++
Technical level of development	+	+	+	+	++	++	++	++

0 not dependent, + dependent, ++very dependent

Table 1: Criteria for lowland typology and their relationship to development parameters (According to Albergel, 1993)

7-1- MORPHO-DYNAMIC PROBLEMS RELATED TO THE NON WATER MANAGEMENT

The greatest concern in the development of the lowlands remains the lack of water control. Of course, the difficulty of controlling water in the lowlands is linked to the quality of the hydro-agricultural facilities and the effective management of water, while knowing how to deal with climatic hazards. The risk of flooding or drying up is a certainty, especially in the current climate of climate change. Occasionally, particular situations can give a low-lying area a torrential character. This is no coincidence, given that the meteorological events of recent years have become increasingly extreme. Indeed, there has been an ever-increasing occurrence of flooding throughout West Africa. It is not surprising, therefore, that the lowlands are more vulnerable to flooding during periods of heavy rainfall. Heavy rain increases the water level in the tributaries of rivers. The water tables then submerge the depressional zones, making the lowland areas unusable until the time of drawdown.

The results of the toposequence analysis illustrate this (Figure 3). An interpolation has been made from the altitudes of observation points collected on the site. This was used to map the spread of water and assess the risk of loss according to the level of submergence. Water flows in a NW-SE direction. In the event of heavy rainfall, the water level rises as a result of normal runoff and runoff from the western slopes. This large quantity of water is difficult to channel using the makeshift infrastructure provided for this purpose. The precarious nature of this infrastructure means that it quickly becomes inoperable. Otherwise, they are very quickly damaged for the most resistant.

The longitudinal relief of the lowland, characterised by gentle slopes and essentially clayey soil, favours the accumulation of water in the centre of the lowland and more widely in the alluvial bed. Under normal conditions, the water is not very turbid and flows slowly, carrying light sediments southwards. However, in the event of exceptional, torrential rainfall,

1the erosive action of water exposes the vulnerability of low-lying areas developed outside dams.

The negative impact of the rainy seasons is just as noticeable on hydraulic structures such as dykes in low-lying areas as on the crops grown. The toposequence analysed allows us to identify areas that can be submerged according to a gradient of gravity. Runoff originates on the slopes, the water level rises and becomes increasingly concentrated towards the centre of the depression, i.e. the minor bed of the lowland, as indicated by the contour lines. The high bunds are more likely to be submerged or to collapse in places. As a result, rice-growing basins located in saturated areas find themselves without ramparts following the partial or total destruction of dikes over a period that can sometimes be relatively long, which constitutes a major risk for plants, in this case rice.

Good aquatic distribution within the rice fields is also determined by the presence or absence of micro-models in the lowland. The most important of these micro-models impart a certain rhythm to the flow regime. The degradation of the earthen bunds and, more generally, of the man-made or natural micro-models (geotope on termite mounds) by the abrasive action of the water thus modifies the texture of the soil and the pedological equilibrium of the lowland. Portions of riparian land very close to the lowland, often used to grow groundnuts and maize, are also affected by the erosive action of water on the banks of the lowland.

7-1-1. Water problems linked to run-off

The intensity of the problem varies according to the regional seasonal rainfall pattern. At Katiola, rainfall is around 1000 mm/year. Heavy flooding is caused by this rainfall coupled with a general north-south slope of the lowland. The synergy of these factors causes young rice plants to be uprooted at the start of the cropping season and the rainy season, or at the end of the rainy season.

In the rainy season, there may often be one or more days between two showers. But it is also possible for showers to last for several days. These successive downpours over several days can submerge and sometimes even suffocate crops planted in the lowest parts of the lowland. Even if some plants, such as rice, can withstand submersion for less than 3 days, the consequences can be catastrophic for plants beyond that period. In addition, the texture of the substrate, depending on whether it is impermeable or slightly permeable, affects the rate of percolation. The lowland thus presents two scenarios of water-related problems linked to run-off. This is because the topography of the site, as a receptacle, favours the convergence of surface water. The first scenario presents a low-lying area with high run-off due to the high water flow, causing young plants to be uprooted. The second scenario presents a submerged, slow-flowing lowland due to the site's adequate levelling on clay soil with low permeability. In such a context, the plants remain submerged for a long time. In both cases, the seedlings are severely affected by uncontrolled run-off.

7-1-2. Water problems linked to run-off

The context here is that of interfluve slopes. The characteristics of runoff always depend on the slopes from which they are recorded. The amount of water flowing down the slopes depends on the amount of rainfall. The slopes are characterised by ochre soils. There are areas of induration in places, developed in the form of sterite. These soils are highly draining, as their infiltration capacity decreases from slightly permeable areas to highly indurated areas. Runoff is at its highest in the roughest areas.

Runoff follows a clearly defined process. Moisture is initially applied from the unsaturated zone and a balance is established between the input and the soil's retention capacity. Then, as rainfall increases, more water is added and the equilibrium is broken. Very quickly, the surplus water runs off towards the lowland, flush with the ground or through gullies and ravines. This infiltration capacity diminishes rapidly as the wintering season progresses.

Runoff is more pronounced on lightly vegetated slopes. In fact, runoff generally predominates on the low-permeability formations of the basement. Run-off is also favoured by the extension into these areas of the surfaces stripped to create tracks. In the various places that form the impluvium, runoff occurs as soon as the wetting conditions are sufficient. This initially sheet-like run-off concentrates locally as local topographical irregularities occur, later creating gullies.

Observation shows sheet flow all along the slope and in the direction of the reservoirs. The slope induces lateral dynamics and leads to lateral runoff. Most of the sites affected are the reservoirs located downstream of the ferruginous slopes to the west of the lowland. Run-off from upstream gradually infiltrates and disappears in sandy topographical segments. However, in the clay segments, the runoff mixes with the drainage.

The runoff highlighted here only concerns a small part of the basin, mainly the area of the collecting depression located to the west of the lowland and its surroundings, and not the majority of the lowland.

7-2- PROBLEMS RELATED TO MESO-RELIEF

7-2-1. The shape of the lowland

The morphology of a lowland is a function of the hydrographic network. This network in turn depends on the overall geomorphological framework, on which the local dynamics of the lowland largely depend. It also determines the form of land use. The size of the rice paddocks depends on the shape of the lowland. At Katiola-Gnomonkaha, these rice pits have a surface area of around 30 m^2 so as to occupy as much space as possible in the lowland and in line with water circulation. This form of spatial occupation follows the geometry of the lowland.

In relation to the geological substratum and the nature of the alterations, the network is either loose and spread out in sandy regions or better organised and denser in clay regions. The lowland studied is irrigated by a main drain from which secondary branches branch off in a fishbone pattern, giving the lowland a more or less elongated shape.

The lateral profile of the lowland is slightly V-shaped, with a relatively flat bed overall. The shape of this lowland makes the rice pits located at the foot of the steep slopes very vulnerable. Generally made up of gravelly material, these segments run off directly into the downstream basins, filling them with coarse material. In addition, the greater the volume of runoff and the longer its duration, the more the colluvium-alluvium spread towards the centre and downstream part of the lowland *(Figure 5)*.

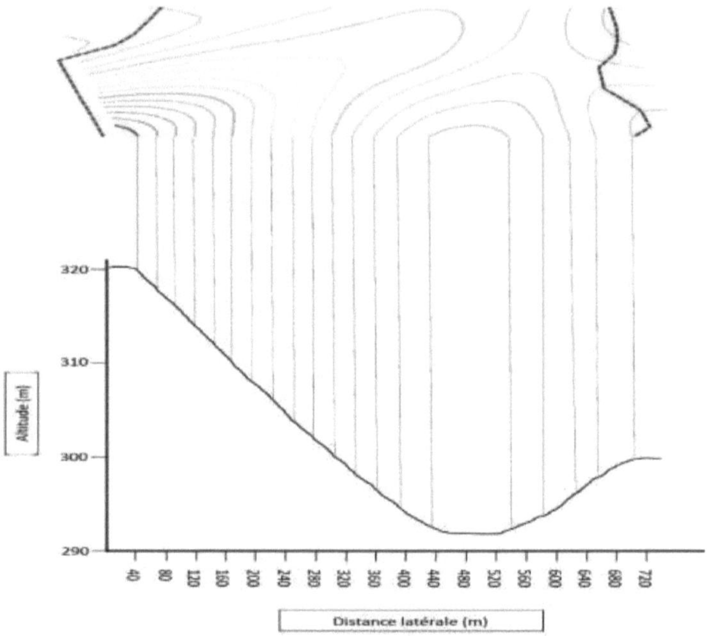

Figure 5: Lateral profile of the lowland

Another problem with the shape lies in its plane of inclination. The bottom is inclined almost horizontally. This results in a relatively slow and fluid flow towards the south. This inclination undeniably facilitates the equilibrium of the water level in the shallows. But the nature of the sediments, which vary from clay to sand, depending on whether you are in the northern or southern part, precipitates a drop in the water level. Because the sandy part accelerates infiltration. So the lowland is constantly balancing the water level by subtracting from the clay part the volume infiltrated in the sandy part.

7-2-2. The soil of the lowland

Diffuse runoff is capable of carrying clay, silt and even sandy particles towards the lowlands. Colluvial-alluvial micro-terraces are then formed at the lowest points of the lowlands, in areas with steep slopes. This type of unit can only be fully observed during the rainy season. Nearby ferruginous soils contaminate sectors of the lowland. These contaminations are materialised by fairly large rust-coloured concretions or patches in the rice paddies. They are reminiscent of the phenomenon of iron toxicity in rice. They are linked to the iron enrichment provided by the sloping groundwater during oblique leaching.

The erosion of the slopes brings veneers of coarse elements to the bottom soil.

This colluvium is a handicap to agriculture, as it sterilises certain plots. The concretions and stains formed in the lowland have a high iron content and expose plants to the risks of iron toxicity mentioned above.

7-2-3. The micro-models of the lowlands

Micro-patterns have been identified in places. They differ according to the nature of their formation. Depending on whether they are of mineral or organic origin, these micro-models are well marked.

Mineral micro-models are generally a kind of residual micro-butt. Mastocline in structure, they have emerged following partial or total desaggregation of the loose part. They can be seen in places in the granitic region towards the western part of the lowland, where the run-off is very marked by projections on the ground towards the downstream end. These micro-models have almost no plant cover due to the hardness of the soil and its ferrous nature. They are not very convex and have gentle slopes. This makes them difficult to identify on the site and helps to camouflage them in the environment. Although in small proportions, these micro-models are an important part of the geological landscape of the site and contribute on a relative scale to the hydric behaviour of the lowland (*Photo 6)*.

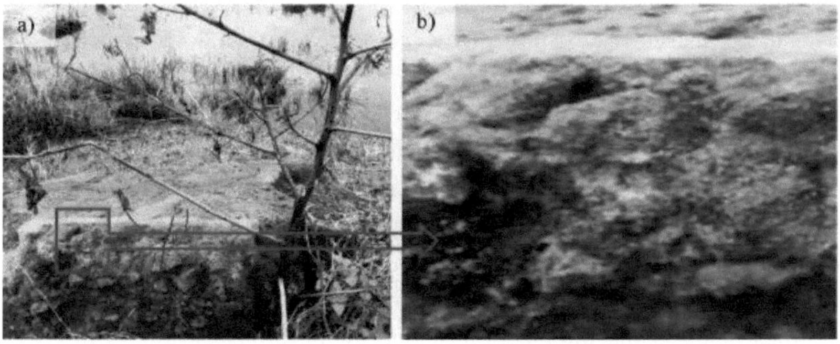

Photographic plate 6: Mastocline structure with brittle texture as it does not facilitate percolation and biological activity

Cliche: BINIE, 2019

The mounds of organic formation are the result of biotic factors. In the lowlands, they correspond to relatively numerous termite mounds. A termite mound is a termite nest, characterised by a mound of hardened earth made from the termites' castings and pierced by numerous openings and channels communicating with the galleries and chambers in the subsoil. These are nitrogen traps. They are easily identifiable on the site thanks to their size. Some of them are up to 2 m high and around 5 m in diameter from the base. They are generally light in colour, although some of them, in this case the smallest, are dark in colour as they are still in the early stages of formation.

The surface structure is in the form of contiguous slits. In the Katiola-Gnomonkaha lowland, termite mounds are more common in the northern part than in the southern part. This spatial disproportion is due to the predominantly clayey texture of the northern part. On their flanks, there is sporadic clear grass cover or branches from nearby creeping vegetation (*Photo plate 7*).

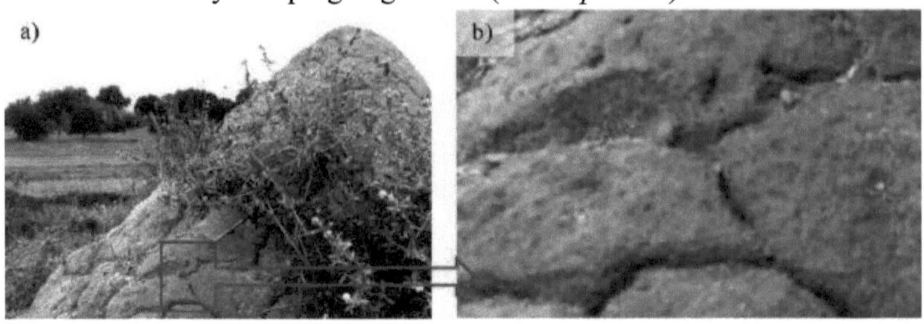

Plate 7: Zoolite-type mega-model with contiguous slotted plates

Cliche: BINIE, 2019

Some isolated termite mounds are still active despite the use of chemicals by rice growers to sanitise cultivated plots. However, many of them have been abandoned and eroded into the lowland. This situation has a certain impact on local agricultural development. Run-off from these termite mounds leads to a build-up of mineral materials in the nearby pits. This addition of soil alters the cleanliness of the soil in the pits, making it more clayey. In the event of a frequent local landslide during the rainy season, a whole portion of the pit, proportional to the termite mound, is damaged. In addition, the presence and evolution of termite mounds in the lowland modifies the distribution of water in the compartments. Sometimes being at the intersection of two or more basins, the termite mound creates a form of depression on the downstream basin and a form of proeminence on the upstream basin. This results in a drop in water flow in the upstream compartment and an increase in flow in the downstream compartment due to the fall. In addition, termite mounds sometimes emerge in the middle of the reservoir. These can be recurrent situations that often lead the rice farmer to abandon the area affected, to the detriment of his potential economic gain. On the site, we observed some abandoned pits despite the dismantling of the termite mounds (*Photo 1*).

Photo 1: Abandoned crate despite the dismantling of a termite mound
Cliche: BINIE, 2019

7-2-4. The risks of abandonment due to lowland grass cover

The geomorphology of the lowlands leads to a diversification of pedological properties. The rice grown there is not entirely uniform across the field. What's more, not all the plots are developed for agricultural purposes in the same way, and so are not maintained with the same care. This leaves room for the rapid growth of weeds, which is a discouraging factor, as it can lead to an increase in the workload for farmers and sometimes to plots being abandoned because of heavy weed cover. Weeding is hard and painful work, especially if the lowland is already sown with rice. In similar cases, the weeding time for a 0.25 ha plot (2500 m^2) varies from 5 to 10 days for one person. This estimate implies a higher production cost in terms of labour in an economically weak environment. Where there is considerable weed cover, rice growers use herbicides (Photo 2). However, these chemicals pose a danger to the plants, even though they offer certain advantages.

Photo 2: Dry grass after easier treatment with herbicide
Cliche: BINIE, 2019

8- ALTERABILITY OF THE HYDRAULIC FUNCTIONING OF THE LOWLAND

The lowlands are the flat bottoms of valleys and small valleys. As such, they form major water drainage routes. The lowlands are also directly influenced by the slopes and peaks of the models, mainly through solid and liquid transport. Hydraulic circulation is organised in a dendritic fashion, resulting in the reduction of flowing water into smaller and smaller branches derived from the main drain. The layout of the branches and the organisation of the confluences show that this is a hierarchical network that requires a great deal of maintenance. However, the geomorphology of the environment gives it a hidden character that could destabilise the hydraulic system.

8-1- AN INCREASE IN THE SEDIMENT LOAD IN RIVERS AND LAKES

The increase in the sediment load of the lowland is due to several actions. These actions relate to the diffusion of water in relation to the topography of the environment (Figure 4). With sedimentary inputs coming from the western slopes and the regions further upstream to the north, the lowland very quickly becomes a receptacle for these source segments. In addition, the dikes of the rice pits act in some ways as rough strips in the lowland. During periods of normal flooding, this configuration of the lowland results in a low flow of excess water. The settling process maximises sediment retention in the shallows. The most depressed areas are highly vulnerable, as the precarious settlements have very little resistance to water pressure. As a result, the risk of crops being destroyed by water is high.

8-1-1. A complex sediment supply

Many factors govern the water supply to the lowland: rain, runoff, rising water tables during part of the year, connections to tributaries, etc. The water movements in the lowland can therefore be superficial or hypodermic. Water movement in the lowland may therefore be superficial or hypodermic. The synergy of these diverse but related factors underpins sediment activity in this area.

Sediment supply fluctuates according to periods of flooding, their duration, and the shape and dynamics of floods, among other factors. This depends on parameters specific to the lowland (size, shape, slope, levelling, etc.) and parameters linked to its environment (relief, vegetation, climate, shape of the catchment area). The timing and height of rainfall, the speed and duration of floods and other parameters determine the load of water in the lowland. They also determine the impact on amenity and crop health.

8-1-2. The formation of alluvial deposits

Because of its relatively flat topography, the circulation of water throughout the lowland is characterised by a network that is not very accentuated. In places, the dispersion of water in the basins describes a constant and continuous flow, forming basins that are flooded in the rainy season and wet in the dry season. However, they may retain some stagnant fagon water until the end of the dry season.

The load of water circulating in the lowland comes from segments located at higher levels. These segments correspond, for example, to the slopes and the tributaries leading to the main drain. The water load is most pronounced at the start of the rainy season. During this period, the soil is drier and more desiccated. As a result, it is rich in fine particles. The first rains, which are particularly heavy, are more likely to erode a dry soil with little plant cover.

The density of the water load is highest upstream and lowest downstream. This is due to the clayey nature of the upstream section. This texture, rich in fine particles, is easily transported by water. It can be identified by the ochre colour of the run-off water that is diffused into the reservoirs. After percolation of surface water bodies, a newly formed clay layer of varying thicknesses, depending on the level of affectivity, covers parts of the casier.

In sandy areas, the water loaded with material forms a deposit characterised by a lighter colour and coarser sediments. This deposit is the result of percolation on fairly permeable, flat ground. The speed and volume of water maintain sinuous traces in the casier to the detriment of the dikes.

The excessive load of water in material has an impact on the layout of the irrigation system and on the texture of the rice paddies. Under pressure, the dikes responsible for channelling the water break. This causes an avalanche of material that weakens or destroys the water pipes. In other places, the channels are blocked due to a large accumulation of sediment where the slopes are shallow. As a result, some paddocks may not be irrigated properly (Photo 3). The rice fields themselves are affected by silting.

Photo 3: Water channel blocked by an overflow of material

Cliche: BINIE, 2019

8-2- CHARACTERISATION OF HYDROMORPHY IN LOWLAND AREAS

The succession of seasons in the course of the year exposes the lowland to alternating periods of submergence and flooding, which in terms of soil chemistry corresponds to a process of oxidation-reduction. Periods of flooding cause leaching, the extent of which depends on the intensity and duration of runoff. As for flooding, it is characterised by desiccation, leading in places to the total drying up of certain parts of the rice-growing basins located in a clay medium. These conditions are linked to fluctuating water levels in the soil in relation to the movement of the water table.

Hydromorphy is a factor that regulates the organisation and intensity of fine particles and therefore the texture and/or properties of the soil. The hydromorphy of a lowland is an essential condition for irrigated rice growing. Lowlands are therefore part of the so-called flood zones. They should therefore be considered as the natural biotope for growing swamp rice.

Hydromorphy varies with the season in the lowland. Submergence during the rainy season creates anaerobic conditions and, conversely, flooding recreates aerobic conditions.

8-2-1. Submergence and risks of iron toxicity

The submersion period is the period during which a large quantity of water runs off the slopes into the lowland. During this phase, a large quantity of material is drained from the often indurated but denuded ferruginous segments and settles in the parts of the lowland where the water stagnates. It has been observed that cases of iron toxicity generally occur around the edges of the lowland, where run-off from the hillsides ends up. In these areas, the water in the reservoir has an ochre colour (Photo 4).

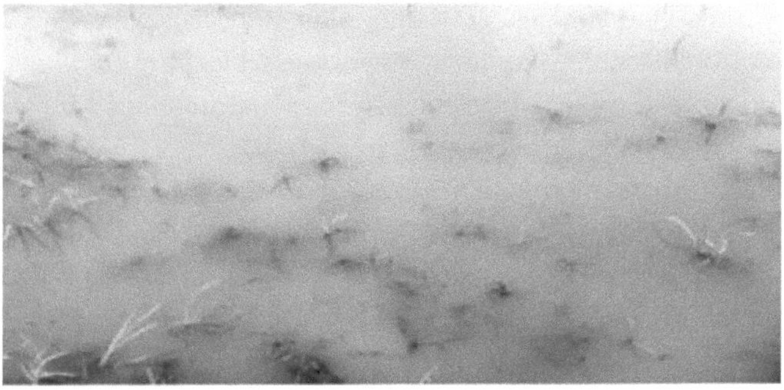

Photo 4: Coloured water from a landfill presenting a risk of iron toxicity
Cliche: BINIE, 2019

Under anaerobic conditions, iron particles are rendered soluble and concentrated in certain areas. Iron toxicity occurs when this concentration of soluble iron particles reaches around 300 ppm. This toxicity is perceptible in rice when the plant is unable to reoxidise its environment through an oxygen supply. As a result, young plants are weakened by the high levels of iron in the environment in which they are transplanted. This weakening often leads to the loss of a large number of plants. The risk of toxicity affects the quality of seedlings. This leads to a change in transplanting dates in relation to flooding risk dates. Submergence with high iron concentrations generally occurs at the start of the rainy season. They give rise to significant and highly coloured runoff. The sowing period is therefore postponed to allow the run-off to dilute the concentration of ferrous elements. When the water circulation system is inefficient, the slower flow results in more residue stagnation in solution. As a result, pockets of residual iron toxicity persist.

8-2-2. Exondation and risks of aluminium toxicity

The generally wet nature of the lowlands limits cases of aluminium toxicity. Admittedly, being in the northern part of Côte d'Ivoire, there are sometimes periods of low humidity. However, the topography of the lowlands favours the presence of a sub-surface water table. However, higher than normal water levels expose the lowland to the risk of alumina toxicity. Alumina toxicity is intoxication after soil mineralisation has resumed. The transition to aerobic conditions leads to re-oxidation. In acid conditions, aluminium becomes soluble and can lead to toxicity, which can be aggravated when rainfall is long in coming. This happens quite frequently in dry regions. The soil becomes increasingly dry and therefore increasingly acidic. Alumina, which is very present in ferruginous environments, is released and conquers the land. In concrete terms, alumina toxicity can be seen in the state of the plants, characterised by root atrophy and a drop in root density. Apart from a few fairly isolated areas, this type of condition is virtually absent from the Katiola-Gnomonkaha lowland. This is because the phenomenon of humification from decomposing litter is underway, blocking alumina production. However, at the end of the dry season, occasional alumina formation can be observed, ga and la. This period is short-lived.

Thus, problems of geomorphological origin in the lowlands can be grouped according to two factors (Table 2): on the one hand, those linked to geomorphological dynamics and, on the other, those linked to hydraulics. Their forms of expression vary in space and time. However, these problems have a considerable impact on farming activities in the lowlands, to the extent that certain basins are sometimes abandoned, crops are waterlogged and cultivation intensifies or migrates to the hillsides. The vulnerability of the lowland is thus becoming apparent (Figure 6).

<table>
<tr><th colspan="2"></th><th>Problems of geomorphological origin</th><th>Identification in the shallows</th><th>Agricultural impact</th></tr>
<tr><td rowspan="10">FIELDS OF APPLICATION</td><td rowspan="7">Geomorphological dynamics</td><td rowspan="4">Lack of water control</td><td>Strong ёсоиlетеп1 from upstream regions</td><td>Plant uprooting</td></tr>
<tr><td>Low infiltration due to clay matëriau</td><td>Submerging plants</td></tr>
<tr><td>Dëbit ёкуё of runoff from dënudës slopes</td><td>Gravelling rice paddies</td></tr>
<tr><td>Runoff constancy on dënudës slopes</td><td>Slope gullying</td></tr>
<tr><td rowspan="3">Irregularity of the meso-relief</td><td>Bottom land reclamation</td><td>Vulnerability of ponds on steep slopes</td></tr>
<tr><td>Coarse veneer</td><td>Stërilisation of plots</td></tr>
<tr><td>Organic and miiK'raux micro-models</td><td>Abandonment of lockers</td></tr>
<tr><td rowspan="2">Hydraulic operation</td><td rowspan="2">Increased sediment load</td><td>Water load according to speed and duration of floods</td><td>Vulnërabilitë de l'amënagement</td></tr>
<tr><td>Relatively flat topography in the lowlands</td><td>Obstruction of irrigation canals</td></tr>
<tr><td>Hydraulic operation</td><td>Alternating hydromorphic conditions</td><td>Slower runoff, accentuated stagnation and risk of Ю\юlё ferrous or ëchoue runoff from hillsides
Flooding and risk of alumina toxics</td><td>Stunting of plants
Root atrophy and a drop in root density</td></tr>
</table>

Table 2: Summary of problems of geomorphological origin in the lowlands

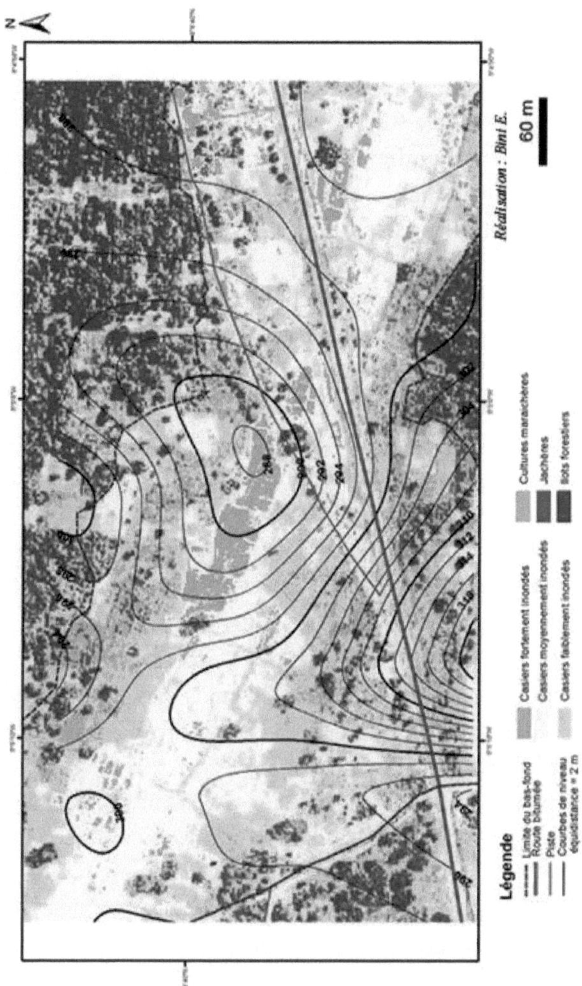

Figure 6: Vulnerability of agricultural issues in the lowland analysis sequence

Source: Google Earth.

CHAPTER 3: LOCAL RESILIENCE TO GEODYNAMIC CONSTRAINTS

Inland valleys are generally perceived as fertile environments which, if properly managed and optimally developed, could provide an opportunity for the development of a rural community and/or greater social comfort for the people who work there. The fact that in some cases these areas are poorly developed is due to a lack of technical mastery, since the many constraints, particularly of a physical nature, remain despite the efforts of the people who use them. The paradox is that we are now in a situation of natural resource scarcity, land impoverishment and economic difficulties, while human pressure on the lowlands is increasing. However, the appropriation and full economic exploitation of these geodynamically constrained environments require a certain level of technical aptitude and financial capacity on the part of the peasantry. Inland valleys combine structural complexity with environmental fragility. Getting the best out of these areas therefore requires the development of local resilient techniques.

9- ADAPTING TO THE GEOMORPHOLOGICAL PROBLEMS OF THE LOWLANDS: AN APPROACH BASED ON OPTIMAL USE OF SPACE

9-1- SPATIO-TEMPORAL ADAPTATION OF SURFACES

Land reclamation on the edge of the lowlands has become an increasingly common practice, aimed at increasing the chances of obtaining additional harvests in a context of climatic insecurity.

The first factor explaining their use is better organisation of the agricultural area. The development of out-of-bottom plots provides good access to the rice-growing basins. In fact, rice growers opt to cultivate adjoining plots as a way of maintaining the perimeter of their farms. The logic of cleaning up the area around the lowland encourages regular maintenance of the plots planted in the lowland. In addition, with a view to controlling run-off, the cultivation of hillsides helps to create breakwaters where they can be created. In this way, the gap between the edges of the lowland and the riverbed itself helps to reduce the speed of run-off on the slopes. This type of arrangement is best suited to areas with sandy or gravelly-sandy soils. This arrangement maximises the amount of stagnant water through efficient percolation.

Crop diversification can be seen as a second factor justifying the development of hillsides and other lowland edges. The agricultural units developed in these areas are essentially market gardening. This form of agriculture is possible and permanent thanks to two major advantages: the accessibility of plots of land and the availability of water resources. Market gardening on hillsides helps to make

the most of a lowland area. The scarcity of the resources needed for agriculture, which requires water and high-quality organic matter, generally calls for integrated management of riparian areas.
Much of the value of the lowland lies in the development of rice growing. The pedological qualities of the site and the availability of water are the main reasons for its attractiveness. However, agricultural activities have gradually spread to the edges of the lowland. The lack of previous data makes it difficult to assess the extent to which this new land is being used. However, testimonies from farmers provide a few snippets of explanation, such as the random nature of the purchase price of rice production, which motivated many farmers to opt for greater financial and food security by diversifying their farming sites. So, in addition to the lowlands, food crops are grown on the marginal areas. However, increasing the size of cultivated plots depends on the operator's financial resources. Areas generally vary between $5m^2$ and $50m^2$. These areas are gradually enlarged to provide additional resources, as explained above.

9-2- EFFORTS TO OPTIMISE GROUNDWATER MANAGEMENT AND RUNNING

Achieving the objective of managing the water table and runoff in the lowlands depends on the stakeholders' understanding of the topography. From a purely technical point of view, the aim is to control water inflows and outflows so as to maintain a sufficient water table for the physiological needs of the rice. It's a constant search for balance that needs to be controlled, both in the rice-growing basins and in the water distribution channels. The fact that the development of the lowland is not of an advanced type limits its ability to withstand periods of flooding. It has all the shortcomings of an archaic infrastructure. So even though raising the water table in the lowland is an advantage, it is still feared by the farmers. The lowland does not have a removable sluice to regulate the flow of water from its point of entry, as is the case in the improved areas under the dam. As a result, it is highly vulnerable to degradation when the flow intensity increases. In the absence of flood calculation tools and modern technical means, farmers enlarge the main drain as they see fit to evacuate floods, despite the risk of oversizing it. In the rainy season, given the increased risk of deterioration, farmers regularly clean out the main drain. This task becomes even more delicate as the force of the water current increases. The ability to maintain an acceptable depth in the drain becomes the main technical concern. Under the pressure of the water, bunds can break, allowing a flow of water that is potentially harmful to young plants. The risk is even greater when the flooding lasts for a long time and the flow reaches a velocity that could cause the plants to be uprooted. To correct these anomalies, farmers resort to makeshift methods. The holes are filled with sand, wooden planks or branches placed side by side (*Photo 5 and Photo 6)*.

Photo 5: Filling a gap using planks Cliche : BINIE, 2019
Photo 6:Filling a gap using sandbags Cliche : BINIE, 2019

When the volume of water is too great, the farmers retreat, letting the water level fall by itself before rebuilding the damaged dikes.

In the dry season, the trend is quite the opposite. With dwindling water reserves, farmers have to resort to empirical techniques to retain soil moisture as much as possible. To do this, they use hoes. These tools are used to loosen the soil. The technique consists of stopping evaporation, which is high at this time of year and tends to dry out the soil, by closing the pores 15 days after the plants have emerged. If necessary, this operation is repeated every 15 days.

As a result, the appearance of the lowland varies with the seasons. The succession of seasons imbues the soil with properties that determine the nature of farming activities. The main factor governing variations in the size of cultivated areas and crops grown is humidity. The greater its extent, particularly in the rainy season, the more spread out the cultivated area. The less it is extended, the more the cultivated area is reduced to the lowest sector of the depression (*Table 3* and Table 4).

Table 4).

Lowland segment	Dry season					
	Geon	Soil	Texture	Land use	Land use	Hydrological regime
	Superior metahedral	Ferruginous non-leaching	Cuirass	Xerophilous grassy vegetation	Not cultivated	Insignificant runoff
	Lower Mdtaedre	Ferruginous, low lye	Sandy-clay	Savanna arboree	Arboriculture	No runoff
	Infraedre	Ferralitic	Sandy-clay	Grassy vegetation	Vegetable growing	Brief flow
	Cataedre	Ferralitic	Clay-loam	Humus	Rice	Brief, intermittent flow

Table 3: Organisation of the lowland segment in the dry season

Lowland segment	Rainy season					
	Gëon	Soil	Texture	Land use	Land use	Hydrological regime
	Mëtaëdre supërieur	Ferruginous non-leaching	Cuirass	Savane arboree	Arboriculture	Intense runoff
	Mëtaëdre inferior	Ferruginous, low lye	Sandy-clay	Grassy vegetation	Vegetable growing	Medium runoff
	ЫIraëëre	Ferralitic	Sandy-clay	Grassy vegetation	Rice	Intermittent flow
	Cataëdre	Ferralitic	Clay-loam	Peat	Rice	Intense flow

Table 4: Organisation of the lowland segment in the rainy season

10-RESILIENCE TO BIOTIC, ABIOTIC AND TECHNICAL-ECONOMIC CONSTRAINTS

10-1- CROP DIVERSIFICATION

One form of resilience developed by local players consists in varying crops. Land use depends on its properties and the seasons. In an area with multiple edaphic characteristics and subject to an evolving climatic calendar, defining a specific crop for the entire lowland would mean a loss of production (*Figure 7).*

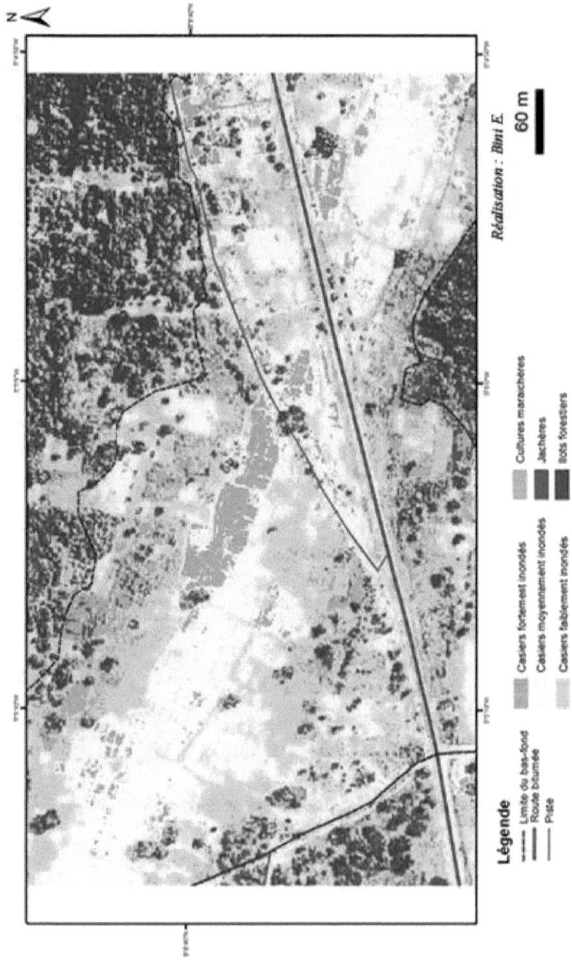

Figure 7: Land cover and land use in the lower analysis sequence

Source: Google Earth.

10-1-1. Distribution of crops by soil

Isolated ferruginous gravelly soils are found at the western edges of the lowland. These soils are partially evolved and transformed by erosion. They are shallow and have a high proportion of stony elements. On these soils, the risk of erosion by run-off is limited by the permeability associated with the presence of pebbles. The heterogeneous, loosely compacted nature of the soil maximises porosity, and therefore infiltration. They are used for tuber crops, notably potatoes and manioc downstream, and maize on the slopes (*Photo 8)*.

Photograph 8: Ferruginous soil marked by gravelly material

Cliche: BINIE, 2019

Clayey-sandy gravelly soils are found on abandoned or less-used drainage routes. These drainage routes are progressively delaminated due to the obstruction and subsequent diffusion of elements from the dismantling of the neighbouring breastworks. The soils in these regions are more clayey at depth, but at the surface they are characterised by the presence of sandy or sandy-silty sediments. Peanuts and beans are grown on these soils (*Photo Plate 9)*.

Photographic plate 9: Sandy-clay soil with anthric properties supporting groundnut cultivation

Cliche: BINIE, 2019

To the south-east of the lowland, sandy-clay soils offer favourable conditions for the development of other types of crop. The texture of these soils encourages farmers to opt for gardening. There are plots of onions, carrots and cabbage (*Photo 10)*.

Plate 10: Sandy-clay soil with anthric properties suitable for growing onions

Cliche: BINIE, 2019

10-1-2. The seasonal distribution of crops

10-1-2.1. Winter crops

Rice remains the main winter crop. The first crop cycle starts as soon as the first rains appear. Rice is either sown directly by planting seeds or transplanted after a few weeks. Sowing generally takes place during the last decade of June. However, climatic uncertainties force farmers to postpone the sowing period until they are certain of a relatively favourable period. For lack of a power tiller, the preparation of the trays for sowing is done by hand.

Nursery sowing is generally carried out by women. Sometimes they come from the rice grower's family, but very often they are women hired on a periodical contract. A day's work costs around 2,000 CFA francs per person. One person usually covers 5 to 10 trays of around 150 m^2 each in a day. Transplanting seedlings has the advantage of requiring a lower water table. In this way, plant growth is less affected in the event of winter disturbances marked by either a drop or temporary interruption in rainfall.

10-1-2.2. Off-season cultivation

Off-season cultivation takes place during the dry period between December and March. It is done on ridges with watering. The lower water level in the lowlands limits the area available for cultivation. What's more, the lack of farming equipment to mechanise irrigation accentuates the difficulties in unfavourable seasons. Farmers therefore turn to the remaining pockets of depression that are still wet. These sites, generally located on the edge of waterholes or close to drains, are the ones that will be able to support off-season market garden crops. These include tomatoes, aubergines and chillies. Early planting allows the plants to make maximum use of the soil's moisture. The soil is often covered with straw or palm fronds to keep it moist.

10-2- ENHANCING PEDOLOGICAL PROPERTIES

The lowland has been farmed for around thirty years. When he started farming, the soil had good fertility. This resulted in excellent production at the end of the season. But over the years of intense farming, the fertility has dropped considerably. Today, a better harvest requires special treatment of the soil. At the start of farming, the lowland had a good yield, but regular cultivation has reduced the soil's natural fertility. To overcome this problem, farmers turned to organic and mineral fertilisers.

10-2-1 Mineral contribution

Fertilising is the main technique for enriching the soil with minerals. It aims to make up for the phosphorus deficit in arable land caused by relentless farming. Farmers also use fertilisers to boost production. This process consists of introducing chemical principles that are immediately useful to vegetation. In addition, renewing the water in the basins and carrying out partial reoxidation also helps to limit the concentration of iron, which is responsible for iron toxicity. However, the lack of appropriate equipment makes this operation difficult and uncertain. Finally, the third method consists of ash amendments. Ash forms a deposit after burning. The ash can also be produced on an isolated site, then collected and poured onto the plots to be cultivated.

10-2-2. Organic input

Soils lose some of their properties when they are located on slopes or in the axis of run-off. One way of maintaining their properties is to enrich them by adding organic matter. To do this, farmers use cattle dung. Sufficient dung is added to the soil from the cattle sheds in special containers. If it is brought in before sowing, the dung is spread over the entire plot. It is then stirred into the soil to homogenise it. In the case of plants that have already developed, the dung is sprinkled all around the plant. Another type of organic input consists of spreading rice straw collected from the threshing floor. They are also mixed with the soil on the plots to be cultivated. After periodic watering, the straw decomposes to form a good humus reserve for the plants. Another technique is to make compost from plant and animal debris, and thus fertiliser. During periods of flooding, the addition of well-decomposed organic matter allows some of the alumina to be complexed.

Another organic input technique is the fallowing of certain parts of the lowland. These are areas that are not cultivated because of their low yield. As a result, they are set aside for a few years so that the vegetation can generate new organic reserves in the soil. When farmers feel that a given area has been depleted by successive harvests, they restore some of its fertility by leaving it fallow. In other places, during this time, they subject these areas to total rest or to cultivation (harrowing, rolling, etc.), the purpose of which is to loosen the soil and better distribute the humus produced by the decomposition of vegetation by bacteria or fungi. Fallowing, although beneficial, is rarely practised in the Katiola-Gnomonkaha lowland.

10-3- THE CONTRIBUTION OF STRUCTURES ECONOMIC AND TECHNICAL SUPPORT

10-3-1 A significant contribution up to 1980

Rice has always played an important role in Cote d'Ivoire's agricultural policy since 1960. Farmers were supplied with agricultural inputs and equipment for cultivating rice by rice development organisations (SATMACI, SODERIZ). Their aim was also to equip and supervise farmers. Financial institutions were also involved. Their main role was to set up an agricultural credit system that enabled farmers to acquire equipment, with subsidies up to a certain level. These institutions also housed accounts and were maintained by the profits from sales. In addition, income from marketing the produce collected was used to offset production costs. This favourable environment lasted until the crisis years. The economic situation in Côte d'Ivoire in the 1980s led the government to withdraw from the production sector, with the dissolution of SODERIZ. The rice-growing situation became even more tense with the political and military crises from 2002 onwards. On 1 August 2012, the ONDR was set up by decree to take charge of rice-growing issues and breathe new life into Ivorian rice-growing. ANADER, for example, provides lowland users with technicians who are sufficiently equipped to provide local support. As for the CNRA, it is working with other organisations on the development of new rice varieties such as "R-Bouake", "8 to 9" and "GT-11". These varieties are being developed with a view to making them more resistant to lowland environmental conditions, and more profitable both economically and ecologically. However, local support is strongly recommended by rice growers, as rice growing is their main or even only activity. However, they very rarely benefit from capacity building through external assistance or periodic training. Many of them are less aware, for example, of the advantages of a social organisation based on mutualism. The same applies to occupational diversity, particularly rice and fish farming.

10-3-2 A laborious mechanism for financial resilience

The dissolution of the state bodies responsible for supervising rice growing in Côte d'Ivoire forced many farmers to give up rice growing. For those who remained in the business for the longest time, this marked the start of a long period of difficulties in finding the financial resources needed to exploit the lowlands. The rice growers then organised themselves into cooperatives in order to pool their efforts around a common goal: revitalising rice-growing activities. In the north-central region of Côte d'Ivoire, CORICENTRE offers rice growers an associative framework whose aim is to unite in order to carry out their activity more cheaply and share the profits. On a practical level, it helps members acquire inputs, plant protection products and power tillers wherever possible. At harvest time, producers bring their produce to the cooperative so that it can be better sold, and the total costs borne by the cooperative are deducted from the revenue. Another means of financing is to turn to private individuals. This practice reduces the farmer's purchasing power, as the tonnage collected is just enough to cover the loans taken out with these traders. The lack of support from microfinance institutions or banks exposes rice growers to abuse by traders. The latter offer them loans that they repay directly to the lender by selling their production. This sale is made to the traders, generally in times of plenty. Thus, in accordance with the law of supply and demand, the large quantity of rice available on the market encourages the trader to reduce the selling price of the rice growers considerably. For their part, rice growers cannot wait any longer and sell in times of shortage, because their creditors are now demanding repayment of their loans. The rice growers therefore fall victim to the vicious techniques used by traders to become the sole resellers in the future and take the lion's share of the profits. In the end, farmers are financially weakened or, in some cases, forced to give up growing rice because they are unable to pay off their debts and finance their farms themselves. The resilience of the peasantry is not uniform in all areas (Table 5). Even if the social aspect were to encourage solidarity in the use of plots, efforts to improve the situation are hampered by a lack of financial and technical resources.

AREAS	LEVEL OF RESILIENCE
Socio-economic integration	
Working group	Student
Management structures	Low
External sources of financing	Low
Technical skills and work capacity	
Hydraulic and agricultural technology	Low
New speculations	Medium
Resilience of the rice-growing activity	Student
Agrarian unity	
Spatial adaptation	Medium
Spatial dynamics	Low

Table 5: Farmers' level of resilience by area

11- SUMMARY OF RESULTS AND MAIN BENEFICIARIES

11-1- SUMMARY OF RESULTS

The ultimate aim of the study is to obtain concrete results that will provide a better understanding of the problems associated with lowland relief. The results presented are of a demonstrative nature. The aim is to obtain elements that will enable us, using GIS, to model how these problems materialise in this type of area (Table 6). The results of this study are also of a factual nature. By observing and analysing the adaptation mechanisms of local populations, we have tried to draw out models that attest to possibilism.

The integrated analysis of the physical and anthropic data interacting in this space has enabled us to draw up a conceptual data model (CDM). This model attempts to make more explicit the perception of the object of the present study by first identifying the different entities present, their attributes and the relationships that exist between these entities, justifying the importance of analysing the environment using a systemic approach.

11-2- MAIN BENEFICIARIES

Applied research is research that aims to put its methods at the service of the people by focusing on their concerns and providing answers. As a result, the main beneficiaries of this study are the farmers who use the lowlands. But there are other beneficiaries as well. State, administrative and university bodies are increasingly interested in scientific support for decision-making. The results we are presenting are intended to make a modest contribution to the rice-growing sector. In addition, the correct interpretation of cartographic and geographic results concerning the vulnerability of lowlands to geomorphological phenomena is essential for accurate decision-making in lowland development.

In addition, the methodology used and the forms of adaptation identified can be reproduced in other national or sub-regional regions that have more or less the same characteristics as those of the environment studied (*Figure 8*).

NATURE OF THE CONSTRAINT	ORIGIN AND CAUSE	PRODUCT EFFECT	PRACTICAL IMPLICATIONS	PEASANT CORRECTION
Predominantly clay (sandy-clay, clay, silt-clay)	Quaternary alluvial deposits	Blockage Drying cracks	High cost of fitting out	Opening / Closing irrigation valves Motorised weeding
Predominance of sand (sandy-clayey, sandy-loamy)	Quaternary alluvial deposits	Permdability of underlying layers	High cost of building agricultural reservoirs	Clay fill Dyke fortification
Heavy load on rivers and lakes	Degradation of the surrounding slopes (deforestation, use of pesticides)	Filling of troughs and shallows with sedimentary materials Pollution by Increase in turbidity as a result of pesticide leaching Morphological instability of river beds Increase in torrentialite	Increased difficulties in developing water intakes for irrigation Scarcity of water at low water levels and decline in quality due to increased pesticide concentrations Difficulties in using/controlling flood flows, which also cause flooding.	Periodic cleaning Reinforcing the walls Passivity / Purification Slowing down of farming activities Reinforcing channels and walls using makeshift materials

Table 6: Table of the systemic model for analysing geomorphological constraints

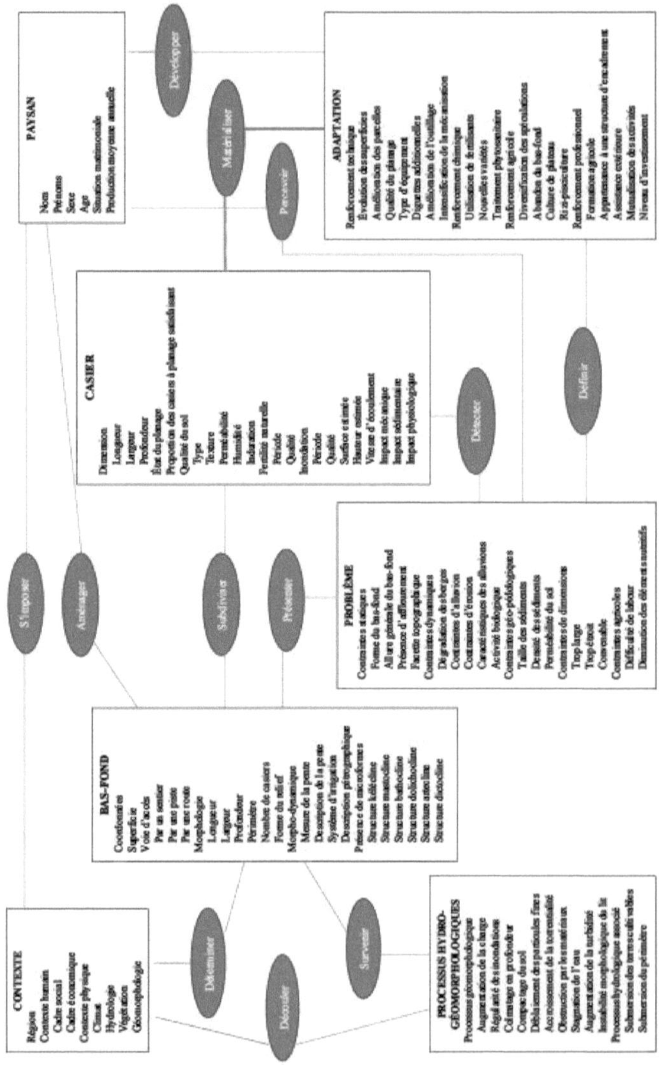

Figure 8: Conceptual model for integrated lowland analysis

CHAPTER 4: DISCUSSIONS

12- THE BENEFITS OF TECHNICAL EXPERTISE FOR USERS OF BOTTOM - FUND

The results obtained in this study are consistent with those of researchers who have looked at other areas. For example, in Benin, Atidegla *et al* (2017, p.15) concluded that rice growers initiate several adaptation strategies, including the conservation of soil moisture during the production season. But resilience is weak and less adaptive.

The problems of geomorphological origin identified in the Katiola-Gnomonkaha lowland are commensurate with the quality of the site's development. In fact, the glaring lack of modern equipment is delaying soil preparation. In a context where the climatic calendar is uncertain and the geomorphological constraints heavy, the low technical level of the farmers exposes them to enormous agricultural difficulties. In order to overcome these difficulties, it is essential that agricultural modernization and the development and use of motorised machinery take root.

For the time being, motorised services are expensive. As a result, most rice farmers are unable to use them, let alone equip them. The low purchasing power of lowland users explains this situation. Agriculture, in all its forms, should be modernised to keep pace with population growth. Otherwise, production would be insufficient. The work we have done confirms the conclusions of Konate L. (1996). In his view, the low level of water technology used by lowland farmers explains the narrowness of cultivated areas and the uniformity of development methods. The same pattern is found everywhere: rice cultivation in the rainy season and market gardening in the off-season (p.169).

The Katiola agglomeration, because of its plateau location, benefits from a relief incorporating several low-lying areas of varying dimensions. This natural opportunity could be fully exploited in terms of agricultural productivity as part of the government's food security policy. One way of achieving this would be to also remove the geomorphological obstacles upstream of development, i.e. during the development of the lowlands. So rice will only yield more in the region once farmers have mastered all the geomorpho-pedological constraints associated with production sites.

13- THE PARTICIPATORY APPROACH TO LOWLAND FARMING

The import policy has resulted in little protection for local producers. An efficient approach should encourage the pooling of all stakeholders involved in lowland farming. This is why more profitable use of the lowlands requires a good knowledge of these environments (Seydou O., 2009, p.40). But acquiring this knowledge as data requires the use of more appropriate tools. Thus, the use of a Geographic Information System is an appropriate tool for more efficient lowland management. Taking this tool into account leads us to the same conclusion as Seydou (op. cit.), who acknowledges that the Geographic Information System (GIS) was used as a tool adapted to the problem and enabling a better approach to decision-making. However, this tool should be used in such a way as to broaden the parameters to be taken into account in the analysis of lowlands. In other words, it must be integrated into a system that takes into account the problems of soil degradation, water redistribution and the imbalance of the hillside cropping system (Konate L., 1996, p.169).

Beyond Katiola, a large-scale mapping of the potential of Côte d'Ivoire's lowlands could help define the actions and responsibilities required to optimise the economic value of these socio-environmental systems. The goal of making Côte d'Ivoire self-sufficient in rice is within reach. To achieve this, the country needs to build a competitive and sustainable food-producing agriculture based on a combination of three factors: high-performance technologies, adequate basic infrastructure and a healthy, well-controlled biophysical environment.

A participatory, multi-stakeholder approach is therefore essential. Insofar as Côte d'Ivoire benefits from a number of advantages, including a climate that is generally favourable to agriculture, a long tradition of rice cultivation by certain rural populations, large areas of rice land that can be cultivated but which are still unexploited or under-exploited, the existence of new varieties and improved, high-performance cultivation techniques, there is a need for good synergy between all the players involved in the production, processing and marketing chain. What's more, every link in the chain needs to be involved in multidisciplinary research, within the framework, of course, of government policy, which has seen an innovation with the creation of a ministry exclusively responsible for rice growing in Côte d'Ivoire on 4 September 2019. This bodes well for the future in terms of improving the management of lowlands to boost rice growing in flood-prone areas.

CONCLUSION

Agriculture is a pillar of the Ivorian economy. It has enabled Côte d'Ivoire to assert itself on the international stage thanks to its many natural assets and its production. But with a fast-growing population, a highly competitive international market and ever-increasing human demands, the need for food is increasing all the time. Rice is one of the most widely consumed foodstuffs in the world. This calls for the entire agricultural architecture to be designed down to the last detail, from biophysical factors through production and consumption parameters to market factors. If we want to achieve national self-sufficiency in rice, or even food security, we need to master all aspects of rice production. This means developing all the lowland sites in the normal way, so that they can be used to grow all the off-season food and vegetable crops. This approach would make it possible to eliminate all the upstream biophysical constraints that have so far been a major factor in the productivity of the country's lowland farming areas, as well as the political, technical and social constraints.

The Katiola-Gnomonkah lowland concentrates almost all the problems faced by rice growers in the Hambol region, not least the lack of control over geomorphological constraints. These constraints, which are both static and hydrodynamic, can be summed up in a complex problem: that of controlling water on sites whose development remains virtually unfinished.

Lowland areas are characterised by specific hydrological dynamics. The rhythm of the seasons, the amount of precipitation and the low relief all play their part in the distribution of the hydrographic network in the lowlands. These factors help to increase their alluvial fill.

The geomorphology of the environment increases the sediment load of the water and imposes micro- and meso-models. This in turn affects the hydraulic system, leading to problems of water control. In the lowlands, the dynamics of the environment impose themselves and make the forms of farming vulnerable. However, in order to maintain their activities, farmers are constantly developing resilient techniques that should be highlighted.

For both rice and flooded crops, lowland soils present agronomic constraints. The extent of these constraints varies according to edaphic properties, climatic variations and the nature of the crops. Farmers are constantly confronted in one way or another with land-related constraints. And despite their best efforts, they rarely manage to overcome them satisfactorily. How can they do this without the help of research and tried and tested technical solutions? This is the crux of the study's reflections, which have enabled us to grasp the problem of farmers' adaptation to geomorphological constraints in lowlands that have been improved and used outside dams, using a holistic approach.

So the impact of geomorphology on lowland productivity is a reality.

BIBLIOGRAPHICAL REFERENCES

ADJANOHOUN Edouard, (1964): *Vegetation des savanes et des rochers decouverts en Cote d'Ivoire Centrale*. O.R.S.T.O.M., PARIS 238p.

ALBERGEL Jean, CASENAVE Alain, RIBSTEIN Pierre, VALENTIN Christian (1992) : *Aridite climatique, aridite edaphique : etude des conditions de I'infiltrabilite en Afrique tropicale seche*, In E. Le Floch, M. Grouzis, A. Cornet, J.-C. Bille, L'aridite : une contrainte au developpement, ORSTOM, pp123-130.

ALBERGEL Jean, LAMACHERE Jean Marie (1993) : *Typologie*, In Mise en valeur agricole des bas-fonds au sahel, J. Albergel et J.M. Lamachere Eds, pp.233248.

ALBERGEL Jean, PEREZ Pascal (1993): *Bas-fonds et riziculture Antananarivo,* 9-14decembre1991 CIRAD, p155-164.

ASSI Kaudjhis Joseph (2008): Crise agricole et reconversion des bas-fonds par la riziculture dans le centre -ouest de la Cote d'Ivoire : enjeux de 1'implication des femmes, *Revue de Geographie Tropicale et d'Environnement*, n° 2, pp. 20-36.

ASSOUMAN Serge Fidele, SYLLA Daouda, KOLI Bi Zueli, (2015): *Exploitation agricole des bas-fonds de Nanbamkaha, dans la plaine alluviale du Bandama (NORD DE LA COTE D'IVOIRE)*, Revue de Geographie de I'Universite de Ouagadougou, N° 04, Sept. 2015, Vol. 2, pp115-133.

ATIDEGLA Seraphin, KOUMASSI Herve Degla, MOUZOU Toyi, HOUSSOU Senan (2017*): Climatic variability and rice production in the Bas-Fond of Dokomey, Benin*, In J. Rech. Sci. Univ. Lome (Togo), pp. 259-276.

BAH Alpha Oumar, SOUMAH Marie-Jeanne, BEAVOGUI Sekou, KOUROUMA Ismael, KABA Aboubacar, DIALLO Mamadou Tolo (2017): *Programme d'amenagement integre des ecosystemes bas-fonds / versants pour une resilience accrue aux changements climatiques,* CTCN, Ministere de 1'Environnement, Republique de Gurnee, 20p.

BERTRAND Georges, (1978): *Le paysage entre la Nature et la Societe*. Rev. Geogr. Pyrenees Sud-Ouest, 49 (21): 239258.

BOUZOU Ibrahim Moussa, MAIGA Oumarou Faran, AMBOUTA Jean-Marie, SARR Benoit, DESCROIX Luc, ADAMOU Mahaman Moustapha (2009): Les consequences geomorphologiques de l'occupation du sol et des changements climatiques dans un bassin-versant rural sahelien, *Secheresse* vol. 20, no. 1, January-February-March 2009, pp.145-152.

BROU Yao Telesphore (2005): *Climat, mutations socio-economiques et paysages en Cote d'Ivoire*, Memoire de synthèse des activites scientifiques presente en vue de 1'obtention de 1'Habilitation a Diriger des Recherches, Universite des Sciences et Technologies de Lille, 226p.

CECCHI Philippe (1992): Les petits barrages du nord de la Cote d'Ivoire a l'heure

du bilan, *L'eau en partage*, pp.261-284.
CECCHI Philippe, WONGBE Alexis Yte, (1992) : L'etude pluridisciplinaire des petits barrages du nord de la Cote d'Ivoire : justification et strategie mise en reuvre, *L'eau enpartage*, pp.17-41.
CHALEARD Jean-Louis (1988): La place des cultures vivrieres dans les systemes de production en agriculture de plantation : le cas du departement d'Agboville (Cote d'ivoire), *Cah. Sci. Hum*. 24 (1), pp.35-49.
CHERON Christian (1991) : *Methodes d'intervention pour l'amenagement des bas-fonds en Guinee*, In Michel Raunet, Bas-fonds et riziculture, pp.437-450.
CHOROWICZ Jean, DEROIN Jean-Paul, (2003): *Bilan Diagnostic, strategies et perspectives de développement département de Katiola*, 53 p., Conseil Generale du Departement de Katiola.
COULIBALY Talnan Jean Honore (2009): *Repartition spatiale, Gestion et Exploitation des eaux souterraines : cas dupartement de Katiola, region des savanes de Cote d'Ivoire,* Sciences de 1'Information Geographique, Universite Paris Est, 141p.
DABIN Bernard et MAIGNIEN Roger (1979) : *Les principaux sols d'Afrique de I'Ouest et leurspotentialites agricoles*, 15 au 19 octobre 1979 au Nigeria tire de Cah. ORSTOM, SER, Pedol, vol. XVII, n O4.1979, pp.235-257.
DANVI Celestin (1995): *L'amenagement des bas-fonds rizicoles*, Ecole inter-etat des techniciens superieurs de l'hydraulique et de l'equipement rural, Ouagadougou, 66p.
DEPIEU M. E., AROUNA A. DOUMBIA Sekou (2017): Analyse diagnostique des systemes de culture en riziculture de bas-fonds a Gagnoa, au centre ouest de la Cote d'Ivoire, *Agronomie Africaine* 29 (1), pp.79-92.
DEVINEAU Jean-Louis, SERPENTIE Georges (1991) : Projet SALT: Paysages vegetaux et systemes agraires au Burkina Faso, In Marcel Pouget Ed, *Caracterisation et suivi en milieux terrestres des régions arides et tropicale*, Paris, Orstom, pp.373-386.
DIAGNE Aliou, KONE Miaman, SYLLA Kalilou, TOURE Ali (2004) : *Politique rizicole et impact de la libéralisation de la filiere riz en Cote d'Ivoire*, "Seminaire de formation sur les NERICA" Abidjan, 23p.
DIRECTION DE LA PROGRAMMATION ET DE LA PROSPECTIVE/MAEP (2001): *Contribution du secteur rural au document de strategie de réduction de la pauvrete (DSRP)*, Direction de la programmation et de la prospective, Ministere de 1'agriculture, de I'elevage et de la peche, Benin, 89p.
FAHO Theophile (1995) : *Enjeux de la lutte antierosive face au dilemme de la demographie galopante et au retrecissement des terres cultivables dans le centre-est du BURKINA,* In E. ROOSE Ed, Environnement humain de l'erosion, Bulletin

reseau Erosion N°15, ORSTOM, pp.44-58
GEOMINES (1982): *Inventaire hydrogeologique applique a I'hydraulique villageoise. Ministere des travaux publics et des transports*. Direction centrale de I'hydraulique, republique de Cote d'Ivoire, Carte de Katiola, Cahier n° 11, 24 pp.
INSTITUT NATIONAL DE LA STATISTIQUE (2000) : Statistique de la population du Recensement General de la Population et de l'Habitat RGPH 1998.
IUSS Working Group WRB (2015): *World reference base for soil resources 2014, Update 2015. International soil classification system for naming soils and developing legends for pedological maps*.
World Soil Resources Report No. 106. FAO, Rome, 216p.
JOUVE Philippe (1992): *Strategies des societes et des Etats*, in E. Le Floch, M. Grouzis, A. Cornet, J.-C. Bille, L'aridite : une contrainte au developpement ORSTOM, pp.467-472.
KAFILATOU Souberou, AGBOSSOU Euloge, OGOUWALE Euloge (2017) : Inventaire et caracterisation des bas-fonds dans le bassin versant de l'Oti au Benin a 1 aide des images Landsat et ASTER DEM, *International Journal of Environment, Agriculture and Biotechnology (IJEAB)*, Vol-2, pp. 1601-1623.
KILLIAN Jean, TEISSIER Jacques (1973) : Methode d'investigation pour l'analyse et le classement des bas-fonds dans quelques régions de l'Afrique de l'Ouest Proposition de classification d'aptitude des terres a la riziculture. Agro. Trop, pp. 156-172.
KOLI Bi Zueli (1999): Potentialites des paysages geomorphologiques et dynamique de l'occupation du sol dans le Centre-Nord ivoirien, In Cahier nantais n°51, *Geographie ivoirienne,* Institut de geographie et d'amenagement regional-Universite de Nantes, pp. 117-129
KOLI Bi Zueli (2014): *Perspectives offered by the use of multivariate analysis and sig in an integrated analysis of geomorphological landscapes,* Revue de Geographie Tropicale et d'Environnement, n° 2, 11p.
KONATE Lamine (1996): *Les bas-fonds dans le reseau hydrographique du fleuve Gambie : cartographie, regime hydrologique, valorisation agricole et typologie,* Universite Cheikh Anta Diop Dakar, These de Doctorat, Geographie physique, 207p.
LAVIGNE Delville, BOUCHER Luc (1996): Les bas-fonds en Afrique tropicale humide strategies paysannes, contraintes agronomiques et amenagements, In Pichot et al eds. *Fertilite du milieu et stratégies paysannes sous les tropiques humides*, proceedings of the international seminar, CIRAD, pp. 148-161.
LESPEZ Laurent, BALLOUCHE Aziz (2009): *Landscapes alongside water: the object, the sense and their trajectories. (Landscapes alongside water: the object, the sense and their trajectories)*. In Bulletin de 1'Association de geographes

frangais, 86e annee, 2009-1 (mars). Paysages au fil de l'eau/ Le patrimoine geomorphologique. pp. 3-11.
MINISTERE DES EAUX ET FORETS (2017): La foret, une priorite du gouvernement, Magazine d'informations du Ministere des Eaux et Forets, N°1 / Octobre 2017, 40p.
NGAMINE Jean (1999): *L'erosion hydrique sur le terroir de Tchanar : une menace pour la durabilite des systemes de culture*, L'influence de 1 homme sur l'erosion, In E. Rosse Ed, Bulletin reseau erosion No 19 Montpellier France, pp.247-257.
NEW PARTNERSHIP FOR AFRICA'S DEVELOPMENT (2005) : *Profil de projet d'investissement bancable : Projet d'amenagement des bas-fonds a participation communautaire,* Appui a la mise en reuvre du NEPAD- PDDAA, Gouvernement de la Republique de Cote d'Ivoire, 28p.
ORSTOM (1987): *Actes du scminaire 21-27 Janvier 1986*, Douala (Cameroun). Editions de I', Collection Colloques et Seminaries Paris.
OUATTARA Seydou, (2009): *Identification des zones favorables aux cultures de bas-fonds dans le bassin du fleuve Bani au Mali et en Cote d'Ivoire*, Institut International d'Ingenierie de l'Eau et de l'Environnement (2IE) juin 2009, 44p.
PEREZ Pascal, SENE Modou (1995) : *Evolution des structures agraires et erosion dans le sud-saloum (SENEGAL*), In E. ROOSE Ed, Environnement humain de l'erosion, Bulletin reseau Erosion N°15, ORSTOM pp.59-68
PIBOT Alain (1996): *Etude de cas : projet de lutte contre l'erosion en zone sahelienne*, In E. Roose Ed, Etat de surface du sol et risques de ruissellement et d'erosion, Bulletin reseau erosion No 16 Montpellier France, pp. 197-206.
POSS Roland, SABATHE Robert (1982): *Etude des aptitudes culturales de la region de Katiola pour les cultures pluviales annuelles, mecanisees, en assolement intensif*, Notice explicative No 94, ORSTOM, PARIS, 34p.
POSS Roland (1982): *Etude morpho-pedologique de la region de Katiola*, Notice explicative No 94, ORSTOM, PARIS, 145p.
RAUNET Michel (1985): Bas-fonds et riziculture en Afrique Approche structurale comparative, In *L'agronomie tropicale*, pp.181-201.
ROOSE Eric (1999): *Evolution historique des strategies de lutte antierosive*, L'influence de l'homme sur l'erosion, In E. Rosse Ed, Bulletin reseau erosion No 19 Montpellier France, pp.11-25.
SERPANTIE Georges, TEZENAS DU MONTCEL Laurent, VALENTIN Christian (1992): *La dynamique des etats de surface d'un territoire agropastoral soudano-sahelien,* In E. Le Floch, M. Grouzis, A. Cornet, J.-C. Bille, L'aridite : une contrainte au developpement ORSTOM, pp.419-447.
SIDIBE Aminata, DJOURARA Hamady, SANOGO Zana Jean Luc, (2004):

Impact de la riziculture de bas-fonds sur la réduction de la pauvrete des femmes rurales de la region de Sikasso: cas des villages de Sola (Cercle de Bougouni) et de Ouatialy (Cercle de Kadiolo), 17p.

SITAPHA Diatta, (1996) *Les sols de bas versant sur granito-gneiss en region centrale de la cote d'ivoire : organisation toposequentielle et spatiale, fonctionnement hydrologique : conséquences pour la riziculture*, These presentee a l'universite de Nancy I,198p.

TOURE Augustin Tiyegbo (1999): Les terroirs de bas-fond de la region de Katiola. Quelle contribution au developpement rural? In Cahier nantais n°51, *Geographie ivoirienne,* Institut de geographie et d'amenagement regional - Universite de Nantes, pp. 151-157.

VIZIER Jean-Francois, DE GIUDICI, RAUNET Michel (1990) : *Etude physico-chimique des sols d'un bas-fond sur les hautes terres de Madagascar : consequences pour la riziculture.* Edits: 30 November 1990, pp.171-177.

ZOMBRE Prosper (1992): Caracterisation morphologique des bas-fonds dans la province du Yatenga, In *programme de recherche en vue de la mise en valeur des bas-fonds au sahel,*

MIX
Papier aus verantwortungsvollen Quellen
Paper from responsible sources
FSC® C105338

Printed by Books on Demand GmbH, Norderstedt / Germany